Stars and Planets

Stars and

The Sierra Club Guide to

Planets

Sky Watching and Direction Finding

by W. S. Kals

Sierra Club Books   San Francisco

Library of Congress Cataloging-in-Publication Data
Kals, W. S.
    Star and planets : the Sierra Club guide to sky watching
  and direction finding / by W.S. Kals.
        p.   cm.
  ISBN 0-87156-671-0
      1. Astronomy—Observers' manuals.   2. Stars—Observers' manuals.
    3. Planets—Observers' manuals.   I. Sierra Club.   II. Title.
    QB63.K25   1990
  523—dc20   90-33711
                                                    CIP

Production by Eileen Max
Cover and book design by Abigail Johnston
Set in ITC Garamond by Classic Typography

Printed in the United States of America on acid-free paper containing
50% recovered preconsumer waste and 7% postconsumer waste
10 9 8 7

Contents

Stars and Planets

I

Finding or Identifying Stars, Constellations, and Planets

1

The Night Sky

The moon and the stars seem attached to the dome of the sky. That half sphere seems to rest on your horizon, the circle where sky and water, or sky and flat land, seem to meet.

Where a building or a hill interrupts the flatness of the land we can, with little effort, imagine an unbroken horizon (Figure 1.1).

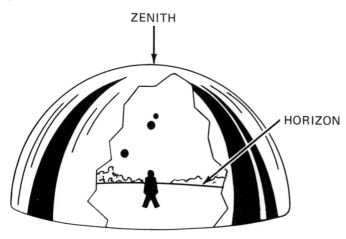

ZENITH

HORIZON

1.1 The dome of the sky, half of an imaginary sphere that rests on your horizon. The moon, stars, and planets appear to be attached to the dome. The point directly above you is your zenith. The line where the sky meets the sea or flat land is your horizon.

We know the dome is an illusion. The stars are not all at the same distance from us. We also know them to be much, much farther away than the moon.

If the Earth stood still we would always see the same stars in the same part of the sky. But looking at the sky in the evening, the middle of the night, and again before dawn, you'll notice that the stars rise, reach a high point in mid-sky, and sink—just like the sun.

We all know that to be another illusion. The dome stands still while the Earth spins around its axis. But we can live with that illusion. For now let's just imagine the stars attached to a whole sphere, rather than a half sphere. One half of the sphere is above the horizon at any given time and place. That half forms the apparent dome of your sky.

If turning on its axis were the only motion of Earth, we'd see the same stars in the same part of the sky at the same time every night. We don't.

Look south around midnight at Christmas time. Anywhere in North America or Europe you'll see the constellation Orion high in the sky. Look six months later, in mid-June. You could look all night. You won't see Orion.

What stars you see, then, depends on the time of night and the season. But that's not all. You'd notice another change in the stars when you travel north or south. In Canada and the northern United States, the Big Dipper—the best known of all star patterns in these areas—will be in the sky every night of the year.

In southern Florida the Dipper will disappear for hours. Stars you have never seen farther north will shine brightly above your southern horizon. In Key West you might even get a glimpse of the Southern Cross.

The problem of finding or identifying stars, then, hinges on this question: What part of the complete star sphere is above the horizon here and now?

One aid for answering that question is a star globe. Like a globe of the world, it should be mounted so you can turn it. It should also let you tilt it so you can study the area around the South Pole.

You could put your sky globe in a box that lets you see only its top half, with the top of the box representing your horizon. You could set such a globe to show only the half of the star sphere now visible where you are. Then the stars would appear in proper relation to your horizon and to one another.

You don't have to get that setting by matching the stars visible in your sky with stars shown on the globe. You can set it ahead of time.

Tilt the globe's axis to reflect your position on Earth, tilting it more when close to the equator, less when you are farther north.

Next spin the globe to the setting for local star time, which depends on the date and local time. Finally turn the box so the proper sides face north, east, south, and west (Figure 1.2).

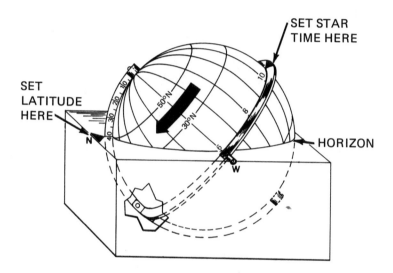

SET STAR TIME HERE

SET LATITUDE HERE

HORIZON

1.2 A star globe, set for your latitude (distance from the equator) and for local star time. It shows the part of the celestial sphere now above your horizon, the top of the box.

Star globes are awkward when you walk the dog or paddle a canoe. They have another shortcoming: You look at them from the outside. That's fine for a globe of the world, but to match the stars easily with those on the globe you should be inside the globe looking out.

Such a star globe, which can be set for any location, date, and time, is a planetarium. In fact, the first planetarium—built long before electric lights had been invented—was a hollow sphere with a door. If you were the Czar of Russia or one of his guests, you'd have climbed aboard in full daylight and closed the door. When your eyes adjusted to the darkness inside, you would have seen the stars: small holes for dim stars, larger ones for brighter ones.

In a modern planetarium it's done by projecting a bright light inside a ball that can be tilted and turned. Pinholes in the ball project the dimmer stars, lenses the brighter ones. You watch the dome from a reclining chair.

It's a great way to watch the stars. Better than outdoors! The view is never spoiled by rain or clouds. You're never cold, and you don't have to swat mosquitos. You can see the winter sky in June or the summer sky in December. A minute later you can compare the two. You can see the early morning stars without losing sleep. And you can see the stars over the North Pole, over Antarctica, or anywhere in between without leaving town.

Unfortunately, you can't change the settings. You have to watch the stars at the place, date, and time set by the operator.

A planetarium seems a good place to learn about finding stars and constellations. But few people have the patience to sit through several hour-long lectures on that topic. So usually a planetarium treats visitors to a trip to one of the planets, a lecture on comets, or a choice of guesses about the Star of Bethlehem.

You probably won't buy your own planetarium, and it's even less portable than a star globe, much less.

A star finder will fit in a briefcase and costs only a few dollars. Some even glow in the dark. You turn a disk for date and time, and the stars that would then be above the horizon show in an oval cutout. The shape of the oval depends on the latitude for which the star finder was designed and works adequately a few degrees north or south of that latitude.

Star finders show the stars as you'd see them by looking straight up. That's not a very comfortable position. So rather than holding the star finder directly overhead, hold it with the word *East* down when you look east, *South* down when you look south, and so on.

Distortions are unavoidable when you try to draw the inside of a half sphere on a flat surface, but since you only look at part of the finder at any one time, its distortion will be less obvious.

Long before the first planetarium or modern star finder, people who wanted to find or identify stars used another method.

They looked for patterns they could recognize in any part of the sky: constellations.

Many of the constellations we use today go back about five thousand years to the Babylonians. They come to us by way of the Greeks and the Arabs.

Constellations are good guides for finding your way around the sky. Once you have located a region from a constellation you have recognized, you can look for more information on a detailed star map of that part of the sky. On star maps that cover only a small area, distortion is no problem.

But there are drawbacks to relying on constellations.

Many constellations are hard to learn. For example, the constellation named after the mythical queen, Cassiopeia, looks at best like a chair. Constellations only partly above the horizon, or partly hidden by clouds, are hard to recognize. Also a bright moon, lights, or sky glow from a nearby town totally overwhelm the fainter stars that make up the patterns of most constellations.

To see the fainter stars, your eyes need to adjust to the darkness, which takes at least fifteen minutes. That's fine in the middle of the night in the mountains, or on an uninhabited island, or on the deck of a yacht far offshore. It's often impossible for people who live in or near towns and cities. Perhaps that accounts for so few people today knowing more than two or three constellations.

This book takes a different approach.

1. You locate stars from a single reference line. That line moves when you travel north or south. (I'll show you how to find it in the next chapter.)

2. You use only the brightest stars for reference. They remain fixed in relation to the reference line and one another.

Using only the brightest stars has advantages: They are visible on bright full-moon nights, without dark adaption, even when you stand under a streetlight. There's a manageable number of them—fewer than two dozen. Unlike less bright stars, they are far enough apart that you are not likely to mistake one for another.

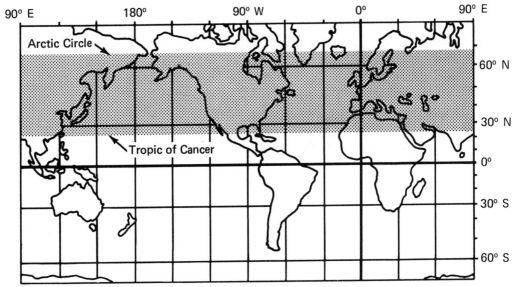

90° E   180°   90° W   0°   90° E

Arctic Circle

60° N

30° N

Tropic of Cancer

0°

30° S

60° S

1.3 Map of the world. The heavy horizontal line labeled 0° is the equator; figures on the right indicate latitude North (above) or South (below) of the equator. The heavy vertical line labeled 0° is the meridian of Greenwich; figures on top show longitude West or East of Greenwich. The shaded part shows the North Temperate Zone; see also Chapter 12.

There's a bonus: The reference line and the bright stars help you find the planets. Because they move among the stars, the planets are never printed on star globes, star finders, or star maps.

If your interest is in learning constellations, starting with the brightest stars is a good method. You'll find more than a dozen prominent constellations directly from the bright stars. These constellations lead you to many less conspicuous ones.

Will my system work anywhere? The answer is a qualified yes. Most bright stars that you'll use for reference are visible in most of the world. You can use the system in Australia or South Africa. But that would make the writing awkward.

Take a simple statement: "These stars rise above the eastern horizon, reach their highest point south of you, then set below the western horizon." (20 words.)

I have just tried to make that statement true everywhere. For example, the stars don't everywhere reach their highest point south of you. They may also be directly overhead, north of you, or below the horizon. Just to give the possibilities, without going into details, takes 73 words.

No reader will want to unravel a snarl of similar sentences. So the first part of this book was written for people who watch the stars between latitudes 30°N and 60°N. It will be just as useful a few degrees north or south of the coverage area, say throughout the North Temperate Zone, 23½°N to 66½°N (Figure 1.3).

The stars don't change suddenly at these borders. So your knowledge gained from Part I will serve you in, say, Alaska, northern Europe, also in Hawaii, Puerto Rico, and the Virgin Islands.

Part II deals in detail with all other areas of the world—from pole to pole. It also deals with applications of your knowledge of the stars, mainly for finding direction from the sky. Methods that work well in one area may not be available or useful in another.

Summary

- The stars seem attached to the dome of the sky, one half of a sphere. The other half of the sphere is below your horizon— invisible—at any given moment.

- What part of the sphere—what stars—you see at a given time and place depends on where you are, the date, and the time of night.

- A star globe, set for these data, would show the stars visible at that time in their proper places. But it's a cumbersome device.

- You could use a star finder, star map, or star atlas. These devices have their own shortcomings, and you have to carry them.

  You could leave them at home if you memorized star patterns, that is constellations.

- Many constellations are hard to remember, difficult to identify in the sky, and often blotted out by the moon or city lights.

- A new approach uses an easily found reference line and concentrates on the brightest stars. It works in a wide area, at any date or time of night.

- The brightest stars lead you to many constellations. The reference line also helps you find or identify the ever-shifting planets.

- In Part II you'll find methods for getting directions from the sky, day or night, anywhere in the world.

2

The
Reference Line

Let's start with a new approach for finding and identifying stars. In Figure 2.1, the dots represent stars. The brighter the star, the more prominent the symbol. All the stars shown are first-magnitude stars, very bright ones. (Don't let the root of the word *magnitude* mislead you. It has nothing to do with the size of a star; it's simply a measure of a star's apparent brightness.)

2.1 Basic star map (also called a strip map). The heavier horizontal line, labeled 0° is the celestial equator, our reference line. Numbers on the right indicate declination (DEC) North or South of the celestial equator. The heavier vertical line, labeled 0ʰ, is the zero-hour circle. Numbers on top indicate right ascension (RA) measured in hours. One hour equals 15°.

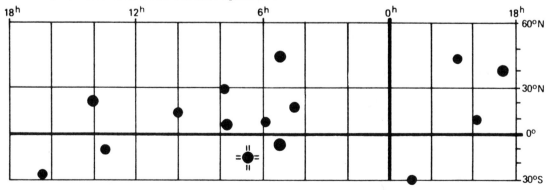

On other star maps in this book you'll find stars of second magnitude and dimmer. You probably won't see stars dimmer than second magnitude when the moon is full. The stars you can just make out under good conditions—your eyes dark-adapted, clear sky, no moon, no city glow—are classed as magnitude five.

You'll read more about magnitude in Chapter 7. For now remember magnitude means brightness, and more prominent symbols—larger dots for example—stand for brighter stars.

Compare the general appearance of the star map and the network of lines with a map of the world such as Figure 1.3. You'll notice in Figure 2.1 a heavy horizontal line labeled 0°. That will be our reference line. It corresponds to the heavy horizontal line with the same label on the map of the world.

In Figure 1.3 you'd call it the equator. For that reason alone you could call our reference line the celestial equator. The equator in the sky is everywhere directly above the equator on Earth. The mast of a vessel steaming along the equator would point to the celestial equator.

The other horizontal lines on the sky map correspond to the parallels of latitude on the map of the world. They are labeled in the same way—in degrees north or south of the equator.

The angular distance from the equator on Earth is called the latitude of a place. Durban is near latitude 30°S, Philadelphia near 40°N. The star near the left bottom corner of the star map is near 30°S, the one closest to the right edge is near 40°N.

Astronomers call the angular distance from the celestial equator declination. The first star then is in declination 30°S, the second in declination 40°N.

You can measure the distance from our reference line, and other angular distances in the sky, with a tool you always carry with you—one of your hands (see Figure 2.2):

- One finger at arm's length from your eye covers about 2° in the sky.

- One hand (four fingers and thumb, your palm and thumb, or fist with thumb outside) at arm's length from your eye covers about 10° in the sky.

- One span—your spread hand from tip of thumb to tip of little finger—at arm's length from your eye covers about 20° in the sky.

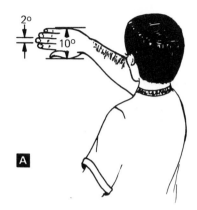

You'll probably agree that four fingers and an edge-on thumb make about five times the width of one finger. You can easily check whether two of your hands equal one of your spans. Just measure with your right hand your left span in "hands."

But you may wonder whether the hands of a child, a woman, and a man cover the same distance in the sky. Generally they do. A child's outstretched arm will be just enough shorter to make its small hand cover about the same angle as a man's hand.

If you already recognize the Big Dipper you can check your own measurements. If you don't, look for its pattern in Figure 2.2. It may be upside down or anything in between. It is visible every night, and all night, everywhere but in the southernmost part of the area.

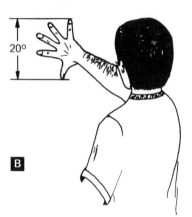

Should you find your own measurements more than a little off, you may add or subtract a little each time you measure. You also could whittle a measuring stick 20° degrees long and marked in quarters each 5° long.

Back to our reference line—the celestial equator. Here are two rules that let you find and trace it in the sky. These rules apply everywhere in the coverage area, and anywhere else in the northern hemisphere.

1. Our reference line (see Figure 2.3) is a half circle on the dome of the sky. It cuts your horizon exactly east of you; it rises to its highest point due south of you; it cuts the horizon exactly west of you.

2. How high this half circle rises in the south depends on your location on Earth in a simple way.

    The angle between the reference line's highest point and the point directly above you (your zenith) is the same as your latitude. In latitude 50°N the South Point of the celestial equator would be 40° below your zenith.

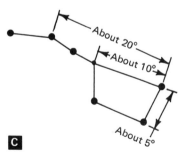

2.2 Measuring angles in the sky. Held at arm's length from the eye, one finger covers 2° in the sky; one hand (palm and thumb) covers 10°; one span (fingers spread) covers 20°.

To locate this reference line you then need to know two things: where is South (which also gives you East and West) and your latitude.

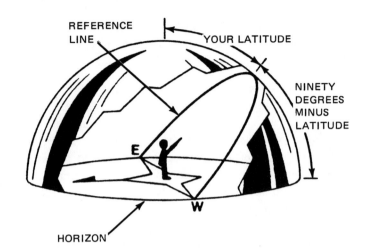

REFERENCE LINE

YOUR LATITUDE

NINETY DEGREES MINUS LATITUDE

E

W

HORIZON

2.3 The celestial equator—our reference line—goes through the points east and west of you on the horizon. It rises highest due South of you. At that point it will be as many degrees below your zenith as you are north of the equator on Earth. In latitude 40°N it will be 40° below your zenith, 50° above your horizon.

At home you probably know where South lies. You don't need great accuracy. Many cities and towns are laid out to make streets or avenues run due south. That's good enough. In other places you may know the direction of a highway, river, or shoreline.

You also may already know an astronomic way for finding North from the North Star, Polaris, or Polestar (Chapter 6). South is directly opposite.

Don't rely on a compass, unless you know that the local error is negligible. Only in few areas does the north end of a compass needle point to true North. By how much and in what direction it misses pointing to North is called variation at sea. On land it's called declination. In the lower 48 states it runs to more than 20°E in the Northwest, 20°W in the Northeast.

The next question is what is your latitude?

You may already know or you can estimate it from the left or right margin of a map. Even from a map of the world such as Figure 1.3. For North America you can get it from Figure 2.4. Example: Winnipeg (Manitoba) is in latitude 50°N. The index of an atlas or encyclopedia also may give the latitude of a city not far from you. For every 70 miles (111 km) you are north of that city, your latitude increases by one degree; it diminishes one degree for the same distance south. It doesn't matter how far east or west of the city you are.

Later you'll see that you can measure your latitude from

Polaris. But for a start we'll trace the celestial equator—our reference line—from your known latitude and known South.

Here is how you do it in four steps (see Figure 2.5):

1. Face South. That brings East in line with your left shoulder, West with your right shoulder.

2. Locate the high point of the equator, due south of you. Its distance from the point directly above you, your zenith, equals your latitude.

2.4   North America. You can find your latitude from the right margin, where every tenth degree is labeled. Each small division is 1°. Example: Charleston, South Carolina, is near latitude 33°N.

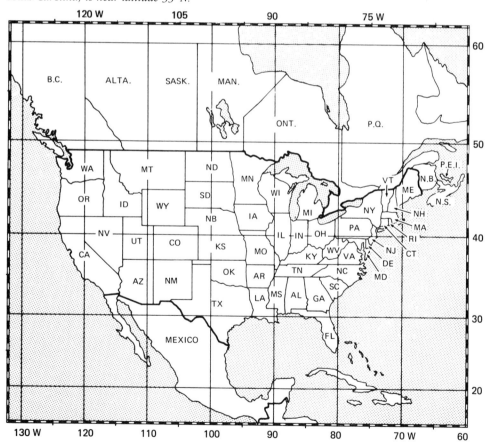

2.5 Sweeping the reference line. Facing south, the father salutes the high point. Then with his arm outstretched, he traces the celestial equator. Illustration by Victor Mays.

For example, in latitude 50°N, in Winnipeg or Frankfurt am Main (Germany), the high point will be 50° below your zenith. You could measure two spans and one hand (or five hands) from the point above you toward the south.

The distance between your zenith and your horizon is by definition 90°; so you could find the high point also by measuring 40° (90 minus 50) upward from the south point on your horizon. You'll probably find that easier.

For example, in latitude 40°N, near Pittsburgh, Madrid, or Beijing, you could measure 40° down from the zenith toward the south. More conveniently, you could measure 50° (90 minus 40) up from the south point on your horizon. In latitude 30°N, near New Orleans or Suez, you'd measure 30° down, or 60° (90 minus 30) up.

3. In your mind, fix the spot just located so you can find it a minute later. Nearby stars or space between stars will help. Then with outstretched right arm give that spot a Roman salute.

4. Still facing South, swing your arm to the right to the western horizon. Next swing your arm to the left. You'll probably bend your arm doing that. At the end of the swing you will be pointing to the horizon due east of you. (You can, of course, perform the salute and the swings with your left arm if that comes more naturally.)

Either arm will have traced the equator in the sky.

You may have guessed it: We are going to line up the equator on a star map, such as Figure 2.1, with the reference line you have just traced. There's only one more problem.

The celestial equator circles the star sphere, just as the familiar equator circles the Earth. As you already know, at any given time and place only one-half of the star sphere is visible, so you'll see only one-half of the celestial equator. The part from East, by way of South, to West will be above the horizon. The other half—from West, by way of North, to East—will be below the horizon and out of sight.

In the next chapter you'll find out which part of the celestial equator is visible at any date and time of night.

Summary

- We can draw a star map using the most popular maps of the world for models. Maps like Figures 1.3 and 2.4 can show only part of the sky or the world. But they include the most important central area.

- On star maps, brighter stars will be shown by more prominent symbols than dimmer ones.

- The horizontal zero line (0°) on the star map, the celestial equator, is our reference line. You will see it as a heavy line on all star maps in this book.

- With one of your hands you can measure angles in the sky. Held at arm's length from the eye, one finger covers 2°; a hand (thumb included) flat or made into a fist covers 10°; a span (fingers spread, from the tip of the little finger to the tip of the thumb) covers 20°. That lets you measure the angular distance between stars.

- Once you know how to find the celestial equator you can also measure a star's declination. This declination corresponds to latitude on Earth. In the sky and on star maps of this area, north declination is above the equator, south declination below. A star in declination 10°N will be one hand above the equator in the sky and on a map. A star in declination 30°S will be three hands below it.

- In this area the celestial equator goes through the point east of you on the horizon. It rises to its highest point due south of you. From there it drops toward the point west of you on the horizon.

- The high point of the celestial equator is as many degrees below your zenith as you are north of the equator on Earth. In latitude 40°N it will be 40° below the point directly above you.

- The point above you is by definition 90° above your horizon. So, in latitude 40°N the highest point will be 90° minus 40°, or 50° above your horizon and due south of you. (You'll usually find it easier to measure from the horizon up than from the zenith down.)

- When you know the direction south and your latitude, you can trace the equator in the sky (Figure 2.5). That line represents one-half of the heavy horizontal line on the star map. The question "which half?" is answered in the next chapter.

3

Star Time

You have just located our reference line, the celestial equator, in your night sky. That lets you align the celestial equator of a star map, such as Figure 2.1, with the sky as it appears at your latitude.

When you worked with the star globe in Figure 1.2, you turned the box so its south side faced south. Then you adjusted the tilt of the globe to match your latitude.

One operation is still missing. You have to spin the globe until its stars match the ones now visible in the sky. The now invisible ones will be hidden in the box.

If you already know a few constellations, matching the globe to the stars won't be a problem. But there is a better way. On the globe (Figure 1.2) you see above the equator the numbers 6, 8, and 10. These figures show hours of star time. The globe is set to local star time 11^h.

On the top of the star map you see similar numbers. To keep the drawing uncluttered, only a few numbers are shown. The lines are two hours apart. You can find the 10^h line to the right of the 12^h line.

Note the point where the 10^h star-time line cuts the celestial equator—the $0°$ horizontal line on the map. That point will be due south of you at the high point of the equator you traced in the sky. That's the very point you greeted with the Roman salute.

The simplest way to find which hour circle is currently south of you is to look it up in a book of tables. Astronomers need great accuracy, but for our naked-eye observations we can make use of the fact that the figures almost repeat year after year. A short table like the one you'll find in Appendix A will do for any year.

Date	Evening Hours						Mid-night	1
	6	7	8	9	10	11		
Jan 5	1	2	3	4	5	6	7	8
13	1½	2½	3½	4½	5½	5½	7½	8½
21	2	3	4	5	6	7	8	9
28	2½	3½	4½	5½	6½	7½	8½	
Feb 5	3	4	5	6	7	8	9	
13	3½	4½	5½	6½	7½	8½		
20	4	5	6	7	8			
28	4½	5½	6½	7½				

3.1 Fragment of Star Time Table (See Appendix A).

The vertical lines would be called meridians of longitude, or just meridians, on a map of the world. But in astronomy the word *longitude* is reserved for a different measurement. That's why on star maps they are called hour circles, a good name since they are usually labeled with hours, not degrees, as are longitude lines. (On our maps they are straight lines, so you may call them hour lines.)

The 0^h is printed darker than the others. It is the origin of the astronomic grid, as the meridian of Greenwich is the origin of geographic longitude. While Greenwich, England, was a fairly recent (1884) political choice, use of the 0^h circles goes back to antiquity and has an astronomic basis.

The 0^h line passes through the vernal equinox, the point where the sun crosses the celestial equator at the beginning of spring. That point was once in the constellation Aries and is still quaintly known as "the First Point of Aries" or Aries for short.

Unlike longitude lines, which are counted west and east of Greenwich, hour lines are numbered straight through. The numbers increase toward the east.

That is most logical. It makes the hour line that bears south of you one hour after the 10^h line the 11^h line. The numbers run to 24. But unlike military time, which confuses many travelers, hour lines present no problem. Except perhaps that the same line may be numbered 24 or 0.

You have seen how the place of a star north or south of the equator is called declination, not latitude as in geography. And you have read that you should not use the term longitude. The term to use for the same measurement in the sky is right ascension, translated straight from Latin.

To spare you reading right ascension over and over let me abbreviate it as astronomical tables do: RA. The star Regulus, RA 10^h08^m, will be close to the 10^h line and bear nearly south when your star time is 10^h.

So you could remember the RA of a star as meaning your star time when it's right above the south point on your horizon. (If you like something fancier than "your star time" you can call it "local sidereal time.")

Astronomers give the RA in hours, minutes, seconds, and four decimals of seconds. I'll give it as 10.1^h, amply accurate for finding stars and planets. (One-tenth of an hour is six minutes.)

You may wonder where star time comes from. It takes the spinning Earth about 23^h56^m to bring you back to where any star will again be south of you. You always thought it took Earth exactly 24 hours. You were quite right if you referred to the sun. The time from one noon—when the sun bears south—to the next noon averages 24 hours.

The Earth also revolves around the sun, once around in one year. So the Earth, having spun you back under the same star, has to turn almost four minutes to bring you back under the sun.

Four minutes a day doesn't sound like much. But it adds up to about one-half hour a week, one hour a fortnight, two hours a month. That explains why a star that was south after the eleven o'clock news is noticeably west of south a month later.

If arithmetic appeals to you, you can calculate star time on the back of a match folder or in your head for star watching or impressing an audience.

We are now ready to put it all together.

• You know how to find the celestial equator in the sky. You also know it's shown on the star map as the heavy, horizon-

Calculating Star Time

- To get approximate star time for **midnight** of any date and year, proceed as follows:

  1. Remember a constant: $4\frac{1}{2}^h$
  2. Add to constant the number of month multiplied by 2^h
  3. Then add $\frac{1}{2}^h$ for every eight days in the month:

for days	add
1–7	$\frac{1}{2}^h$
8–15	1
16–23	$1\frac{1}{2}$
24–31	2

  Examples:

March 7, midnight			**March 22, midnight**	
Constant	$4\frac{1}{2}^h$		Constant	$4\frac{1}{2}^h$
3rd month × 2	6		3rd month × 2	6
7th day	$\frac{1}{2}$		22nd day	$1\frac{1}{2}$
Star time is	11^h		Star time is	12^h

- For **other times,** find star time at midnight then subtract hours before midnight, add hours after midnight:

  Examples:

March 7, 11 PM			**March 22, 5 AM**	
Star time at midnight	11^h		Star time at midnight	12^h
Hours before midnight	-1		Hours after midnight	$+5$
Star time is	10^h		Star time is	17^h

- When star time comes out greater than 24^h, subtract 24^h.

  Example:
  Calculate star time for October 6, at 1 AM local time. You'll get 26^h; use star time 2^h.

---

tal $0°$ line. You also know two ways—table or calculation—for finding star time.

- Say the star time is 12^h. The hour line labeled 12 on the star map will start at the point south of you on the horizon, and it will point straight toward the point directly above you.

- The hour line labeled 6^h—which had been due south six hours earlier—will slope in your western sky. Its zero-declination point, where it crosses the equator on the map, will be due west of you on the horizon.

- The hour line labeled 18^h—which will be due south six hours from now—will slope in your eastern sky. Its zero-declination point, where it crosses the equator on the map, will be due east of you on the horizon.

- Imagine the star map bowed over your sky, the equator on the map touching the celestial equator you traced in the sky. At this time the 12^h line on the map should be due south of you.

- The stars on the map will correspond to the brightest stars in the sky. Stars north of the equator (declination N) will be above the line you traced in the sky; stars south of the equator (declination S) will be below that line in the sky.

- A star of RA 12^h (on the 12^h line on the map) will be south of you, whatever its declination.

- A star on the celestial equator at RA 6^h (on the 6^h line) would just be setting due west.

- A star on the celestial equator at RA 18^h (on the 18^h line) would just be rising due east. All other stars on the star map will be in proper relation with each other.

Now you can see why accuracy in finding South, your latitude, and local star time doesn't matter much. There are only so many bright stars visible at the same time. These stars themselves will correct your errors. Just match the stars on the map with the brightest stars in the sky.

The star map in Figure 2.1 shows fifteen of the brightest stars. Twelve of them are within declination 30°S and 30°N. In that area distortion of the chart is negligible. One hour of RA on the equator equals 15°. Within 30° of the equator—north or south—you can still measure it as one-and-one-half hands.

The remaining three stars on the map are within 15° of declination 30°N. You can still measure declination with your hand, but the hour circles begin to crowd together as you can see on the sketch of the star globe (Figure 1.2).

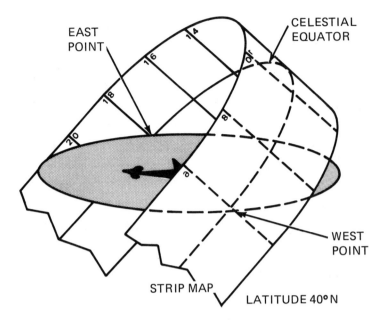

3.2 A star map such as Figure 2.1 is bowed over your horizon. The celestial equator of the map lies over the line traced with your arm.

There are no first-magnitude stars farther north.

The fifteen stars shown on the map are the only first-magnitude stars you'll ever see in most of the coverage area.

Often you'll see one or several bright, even very bright, stars in the sky that are not on the map. They may look like stars, but they're planets. They'll always be within three hands of the equator. You'll read about them in Chapters 7 and 8.

Only near the southern border of the coverage area will you see the remaining six brightest stars. There they will always be, low above your southern horizon. As you travel south, the bowed strip (Figure 3.2) that pivots around the east and west points on your horizon tilts farther back. The six stars—all well south of the bottom of the strip map—will eventually be lifted above your horizon.

In latitude 60°N the bottom of the bowed strip chart just touches your southern horizon. As you travel south, the celestial equator and with it the strip map rise higher and higher. When—traveling south—you reach latitude 30°N, the top edge of the strip at its highest point will be directly above you.

All that is difficult to visualize. It may help to refer again to the star globe (Figure 1.2).

- Our reference line is represented on the globe by the equatorial ring, which pivots at the east and west points.

- The strip map (Figure 3.2) stands for the central zone, north and south of the equator on the globe. You pull the strip until your star time—from date and time—is due south. That's like turning the globe until the correct hour circle faces south.

    I have taken only one liberty. I have replaced the central bulge of the globe with a cylindrical strip. In this central area of the globe that causes only a small distortion.

If you care to examine the tilted strip more closely, you can read from it some sky facts you may already have observed.

- A star on the celestial equator rises due East, six hours later bears South and sets due West another six hours later. Just like the sun at the beginning of spring and fall, when everywhere on Earth it is twelve hours above the horizon.

- A star south of the celestial equator rises south of east, takes less than six hours to reach south, and sets south of west less than six hours later. Just like the sun in winter.

- A star north of the celestial equator rises north of west, takes more than six hours to reach south, and sets more than six hours later north of west. Just like the sun in summer.

In the next chapter you'll see that identifying the brightest stars is easier than the last two chapters may have led you to believe. They are the only hard ones in this book. To help you digest them, a step-by-step review of fitting the star map to the sky of the moment follows.

Summary

- The vertical lines on the star map—corresponding to the meridians of longitude on a map of the world—are called hour circles or hour lines. They are numbered eastward from 0^h to 24^h.

- What we call the longitude of a place on Earth is called the right ascension (RA) of a star. A star in right ascension 10^h will be shown on the 10^h line on the star map.

- Your star time (local sidereal time) tells which hour circle is due south of you at a given date and time. When your local star time is 10^h, the line labeled 10^h will rise straight from the South Point on your horizon to your zenith. A star in RA 10^h will be due South of you.

- You can find approximate star time from standard time, from a table, by calculation, or from the stars themselves.

- The point on the celestial equator and on the hour circle labeled six hours earlier than the present star time (the 4^h circle in the example) will be due west of you on the horizon.

- The point on the celestial equator and on the hour circle labeled six hours later than the present star time (the 16^h circle in the example) will be due east of you on the horizon.

- Align the star map so that its horizontal line, the celestial equator, falls on the celestial equator you traced in the sky. Place the hour circle that corresponds to your star time due South.

  The stars on the map will then correspond to the stars in the sky. (You may have to tug the map a little left or right to correct for the difference between standard time and your local time, and other errors.)

Review of the Method

Know the standard time (subtract one hour from daylight saving time), your approximate latitude, and where South lies, then proceed as follows:

1. Find present star time using Appendix A.
2. Subtract your latitude from $90°$.
3. Measure that angle with your hand upward from the South Point on the horizon.

4. Through the point just found trace the celestial equator (Figure 2.5).
5. Imagine one-half of the equator on the map lying on the line you just traced in the sky.
6. Slide the strip sideways along that line so that the hour line labeled with the present star time is due South. For star time between 6^h and 18^h use Figure 4.1; for all other times use Figure 4.2.
7. Match star symbols on map with brightest stars in the sky. That match corrects for errors in star time, latitude, and direction of South.

4

The 21 Brightest Stars

In this chapter I'll formally introduce you to all twenty-one first-magnitude stars.

You already have a nodding acquaintance with many of them. Twelve lie within 30° north and south of the celestial equator. Three are not much farther north.

That's the fifteen stars shown on the strip map (Figure 2.1). There they were just dots. In Figure 4.1 I group fourteen of them into three patterns. You might call these patterns super-constellations.

All these stars have names, strange names. You may need some help in saying them. Probably you don't want to struggle with complicated codes and symbols such as dictionaries use. One has more than seventy, just for the vowels!

I'll make it as simple as possible. First clue: The stressed syllable is printed in capital letters.

This is the key for the vowel sounds:

a	as in fat	ay	as in fate
e	as in wet	ee	as in feet
i	as in it	eye	as in ice
o	as in odd	oh	as in go
uh	as in up	you	as in use
	oo as in food		

Notice that all the sounds in the left column are short. In the right column they're long and like the names of the letters a, e, i, o, and u.

I'll also use "uh" for the neutral sound we use in unstressed syllables in words like sofa, kitten, pencil, lemon, and circus. Followed by the letter R it will also do for the sounds in earn, irk, and urn. You won't need a special symbol for "ar"; sound it as in arm.

I respell star and constellation names the first time they're mentioned and in some lists. To save you finding this page each time you are in doubt, the key will be repeated in Appendix F.

This simplified scheme will get you close to the preferred pronunciation. Several star names have more than one accepted way of sounding them. Betelgeuse, wins the prize for that. It is usually spelled Betelgeuse. BET-el-guhrz is one accepted sounding. Some insist the last vowel should sound like the French *deux*. Others want a German umlaut as in König. Germans pronounce it Bet-eye-GOIS-uh. Sailors long ago solved the problem. They call the star BEETLE-juice.

Now to the superconstellations in Figure 4.1.

THE HEXAGON: Near 6ʰ, a bit north of the equator, the star Betelgeuse (BET-el-guhrz) is surrounded by a ragged hexagon of first-magnitude stars. That hexagon is about six hands high, five wide.

Clockwise from the top, the stars around Betelgeuse are Capella (ka-PELL-uh), Aldebaran (al-DEB-er-an), Rigel (REYE-juhl), Sirius (SEE-ree-uhs), Procyon (PROH-see-uhn), and Pollux (POL-luhks).

I have made up a silly jingle to help me remember these stars in the proper order. Think of a difficult ship's captain. He wants to see the rigging of his ship polished. On other ships it is painted, slushed with oil, or left to rust. The first mate—a foreigner—reports:

"Captain, all de rigging seems properly polished."

The underlined letters stand for Capella, Aldebaran, Rigel, Sirius, Procyon, and Pollux.

THE DOUBLE TRIANGLE: Centered on RA 13ʰ are two triangles back to back. Both triangles straddle the celestial equator. Their short

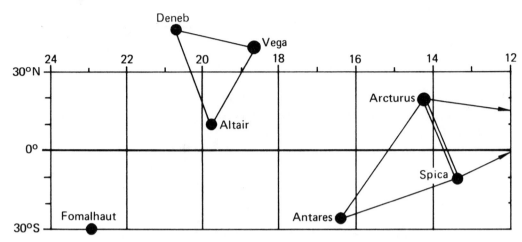

4.1   *Star map for star time 6ʰ–18ʰ, showing names of stars and supercon-stellations Hexagon, Double Triangle, and Triangle.*

sides are about three hands, their long ones between five and six hands.

I think of that figure as two triangles, rather than one with an extra star thrown in. That helps me recognize the Siamese-twins figure when part of it is below the horizon.

Naming the stars as though they were all in one figure, clockwise from right on the map, you have Regulus (REG-you-luhs), Spica (SPEY-kuh), Antares (an-TAY-res), and Arcturus (arc-TOO-ruhs or arc-TYOU-ruhs). Spica south of the equator and Arcturus north of it form the side where the triangles are joined.

To remember these stars in the order given I think of the cook of the ship with the polished rigging. He's making soup for the captain. It contains:

Regular spices and arsenic

Decoded gives Regulus, Spica, Antares, and Arcturus.

THE TRIANGLE: Near RA 20ʰ, about three hands on a side, the Triangle is entirely north of the celestial equator. With the shortest side 2½ hands, the longest 3½, it just misses being an equilateral triangle.

Clockwise from top right—and by chance in order of bright-

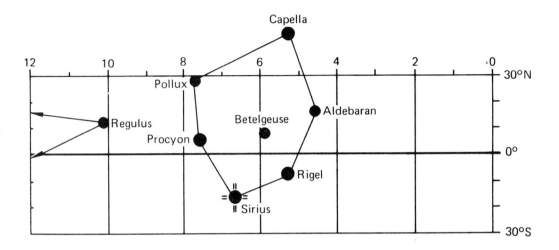

ness—you have Vega (VEE-guh), Altair (al-TAYR), and Deneb (DEN-eb).

Here's how I remember them: The cook is in court accused of doing in the dear captain. The prosecutor proposed to prove that the cook poisoned the captain's vegetables. "No, no, not true," screamed the cook. (After all, it was the soup!) A newspaper headlined the story:

Vegetable Alteration Denied.

Decoded: Vega, Altair, and Deneb.

ONE LONER: A first-magnitude star near 23^h doesn't fit into a superconstellation. It will be south of you when star time is 23^h; since it's near declination 30°S it will at that time be three hands below the high point of the celestial equator you trace in the sky.

In latitude 40°N, for example, the high point of the equator is five hands above your south horizon. There this star will be about two hands above your South Point. Presently you'll get help for finding that star when it is not due South.

The star's name is Fomalhaut (FOH-mal-hoht). According to different authorities the last syllable is also pronounced -hoh, -oh, and -ot.

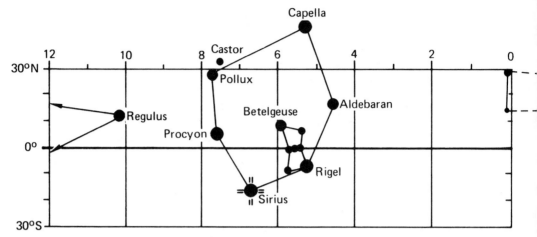

4.2   Star map for star time 0ʰ–6ʰand 18ʰ–24ʰ, with the constellation Orion, the Square of Pegasus, and Castor (the twin of Pollux) added.

We have worked our way from the right edge of the star-map to the left edge. Like a similar map of the world our strip is supposed to be a closed ring in which the left and right edges meet. But you can't print it that way.

Sometimes the stars you see in the sky will be partly on the left half of Figure 4.1, partly on the right. (I'm not talking about the unavoidable break where a book closes.) For example, at local star time 10^h, our Double Triangle will be mostly left when you face south, with the whole Hexagon on your right. No problem with Figure 4.1.

But at star time 0^h, which is the same as 24^h, you'd have our triangle on your right when you face south but on the left on the map. The part of the Hexagon that's still above the horizon will be on your left but right on the map. Awkward! The problem disappears when you use Figure 4.2.

I have added the constellation Orion (oh-REYE-uhn) inside the Hexagon. It has two first-magnitude stars, Betelgeuse and Rigel.

Many people already know Orion, but many think of it as a constellation belonging to the northern hemisphere. Not so. It straddles the equator, which makes it visible all the way to Antarctica.

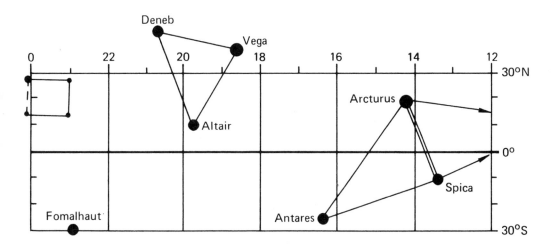

Here Orion is visible for many hours of the night only from September through March. In, say, latitude 40°N in mid-August it rises only about one hour before dawn. In mid-April it sets only about an hour after it's completely dark.

If you don't already know Orion, by all means look for it. It's one of the brightest constellations and its pattern is easy to remember.

It's also useful in naked-eye astronomy.

The three stars that form the belt of Orion are virtually on the celestial equator. That lets you check the sweep of our reference line whenever Orion is in the sky.

The belt points left to Sirius, the brightest of all stars; to the right it points to Aldebaran. Both these stars in the Hexagon are less than three hands away. Rigel, also in the Hexagon, is on the side of the belt opposite Betelgeuse. The distance between the two stars is about two hands, with the belt about halfway between them. Pollux is roughly at the opposite end of the line between Rigel and Betelgeuse, about 3½ hands away.

When your star time is 5½ʰ you don't need to know your latitude or where South is. Orion will be due South of you and its belt will be the high point of the celestial equator.

In Figure 4.2, straddling the gutter—the blank space between left and right pages—I have added four not very bright stars. They form the Square of Pegasus. Our main interest in them here is that the right edge of the square just about points to the orphan star Fomalhaut, two spans from the nearest corner. There are no other bright stars anywhere near it.

The left edge of the square is virtually on the 0^h line. When that edge is due South of you, star time is 0^h (or 24^h).

I have added one more star to the map: Castor (KAS-tor), the twin of Pollux in the Hexagon. Castor, about two fingers from its twin, just misses being of first magnitude.

Figure 4.2 shows all the first-magnitude stars within its area. If you see there a starlike object brighter than Castor, it is not a star but a planet.

You'll read more about the planets in Chapters 7 and 8. For now just a rough definition. Stars glow by their own light, like our sun. They keep station with all other stars, so they are called fixed stars.

The planets—Earth is one of them—circle the sun and are seen by reflected sunlight. Mars, for example, may be quite close to Pollux in the sky at some time, later it'll be more than a hand from that star.

If you live north of about latitude 34°N, the fifteen first-magnitude stars you have now met are the only ones you'll ever see in the sky.

The 34°N line runs roughly through Los Angeles, Phoenix, Atlanta, Casablanca, Beirut, and Shanghai. You can find whether you are north or south of that line from Figure 2.4 for North America, from Figure 1.3 for the rest of the world. All of Europe is north of that line.

South of that line six more first-magnitude stars appear in your sky. They show up in the gap between the bottom of the strip map, tilted for your latitude, and your southern horizon (Figure 4.3).

The first of these stars to appear when you travel south is Canopus (kuh-NOH-puhs). Theoretically it should appear on your southern horizon near latitude 37°N, but near the horizon all stars are dimmed.

A familiar effect. You can't look straight at the sun when it's high in the sky; but at sunrise and sunset its light rays, slanting

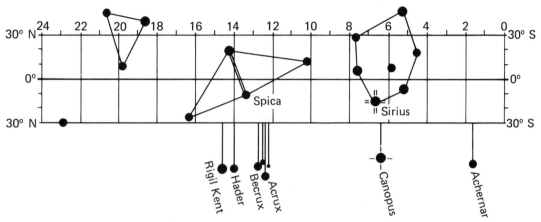

4.3   The brightest southern stars. The six first-magnitude stars visible only in the southernmost part of the coverage area have been added to the basic strip, along with two dimmer stars to complete the Southern Cross.

through the densest part of the atmosphere, are weakened. At these times you can watch the sun quite comfortably.

Canopus, being the second brightest star will appear as a first-magnitude star, brighter than Castor, in about latitude 34°N. Finding it is easy. Being by coincidence almost on the same hour circle as Sirius, Canopus will be near your South Point on the horizon when Sirius bears south (star time about 6½ʰ).

In latitude 25°N, at Key West, for example, Canopus will then be 12°—a little more than a hand—above the horizon. It will have risen 3½ hours earlier, about three hands to the east of South; it will set about 3½ hours later, three hands west of South.

Just as you can find Canopus from Sirius you can find four more of these six southern stars from Spica. When Spica bears south (star time about 13½ʰ) these stars will be low on your southern horizon and close together (Figure 4.4).

About one hour before Spica bears south, Acrux (AK-ruhks) and Becrux (Bay-kruhks) will be due South of you. Star time then is about 12½ʰ. They're two of the four stars that form the Southern Cross in the constellation Crux.

About one hour after Spica bears south, Hadar (HA-dar) and

35

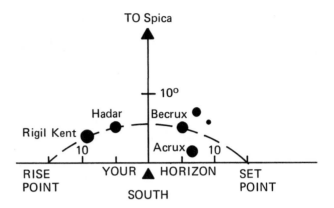

4.4   *Track of southern first-magnitude stars seen from latitude 25°N at star time 13ʰ, when Spica bears south.*

Rigil Kent (REYE-juhl kent)—in that order—will pass due South of you. Star time then is about 14½ʰ.

That leaves only one more southern first-magnitude star: Achernar (AY-ker-nar). It's about as far south of the celestial equator as the top star of the Southern Cross, and so describes here the same small arc low above your southern horizon.

It reaches its highest point when your star time is about 1½ʰ. In latitude 25°N it will be about four fingers above the horizon. That's about 2½ hours after Fomalhaut and the western edge of the Square of Pegasus have been due South of you.

The six southern stars, the twelve within 30° of the celestial equator, and the three not far north of that area make twenty-one stars. That's all the first-magnitude stars ever visible anywhere in the world.

To find first-magnitude stars and the superconstellations they form, a night of full or almost full moon is good. The moonlight will wash out dim stars, leaving you only with bright ones. One difficulty: You may not be able to spot the Square of Pegasus.

Only one half of the strip map is above the horizon at any time. So don't expect to see, for example, Sirius and Altair high in the sky at the same time.

If you do see one or more very bright stars in the area of the strip, not shown in Figures 4.1 and 4.2, you are looking at planets. You'll never see a planet or the moon outside that area.

In the next chapter we'll look at the constellations you can find from the first-magnitude stars.

Summary

- First-magnitude stars are the brightest stars, stars that appear to the human eye brighter than Castor, the twin of Pollux.

- Within a strip of 30° north and south of the celestial equator, there are only twelve first-magnitude stars. Not far north of that area of the sky are three more.

- We can arrange fourteen of these stars into three superconstellations.

    The Hexagon around Betelgeuse in RA 6^h is formed by Capella, Aldebaran, Rigel, Sirius, Procyon, and Pollux.

    The Double Triangle centered near RA 14^h straddles the equator and is made up of Regulus, Spica, Antares, and Arcturus.

    The Triangle, near 20^h and entirely north of the equator, is made up of Vega, Altair, and Deneb.

    The remaining star, Fomalhaut, near RA 23^h at the southern edge of the strip map, is best found from the right (western) edge of the Square of Pegasus.

- These are the only stars that appear as first magnitude north of about latitude 34°N. South of that latitude six more such stars may become visible at times near the southern horizon.

- Canopus will be the highest south of you when Sirius is also about South.

- Four more first-magnitude stars appear when Spica is south. They are—from East to West, which is also the order of their appearance when rising—Acrux, Becrux, Hadar, and Rigil Kent.

- Archernar, the last of the southern first-magnitude stars, will be south of you when star time is about 1½^h. It follows about the same path as the top star of the Southern Cross.

- These twenty-one are the only first-magnitude stars. Planets, which circulate in the area of the strip map, often look like first-magnitude stars, and often much brighter.

These are the 21 brightest (first magnitude) stars. RA indicates star time when a star is highest in the sky and due South (in the northern hemisphere). The stars south of the strip map are visible only south of latitude 35°N.

Brightest Stars Within or North of Strip Map

Star Name	RA[h]	Superconstellation	Notes
Aldebaran	4½	Hexagon	
Altair	19¾	Triangle	
Antares	16½	Double Triangle	
Arcturus	14¼	Double Triangle	
Betelgeuse	6	Center of Hexagon	
Capella	5¼	Hexagon	North of Strip Map
Deneb	20¾	Triangle	North of Strip Map
Fomalhaut	23	*none*	Find from Square of Pegasus
Pollux	7¾	Hexagon	Castor nearby
Procyon	7¾	Hexagon	
Regulus	10	Double Triangle	
Rigel	5¼	Hexagon	
Sirius	6¾	Hexagon	
Spica	13½	Double Triangle	
Vega	18½	Triangle	North of Strip Map

Brightest Stars South of Strip Map

Star Name	RA[h]	Help for Finding
Achernar	1½	Highest 2½ hours after Fomalhaut
Acrux	12½	Highest about 1 hour before Spica
Becrux	12¾	Highest about 1 hour before Spica
Canopus	6½	Highest about ½ hour before Sirius
Hadar	14	Highest about ½ hour after Spica
Rigil Kent	14¾	Highest about 1 hour after Spica

5

Constellations

Everybody knows a constellation is a pattern of stars. If pressed for a more accurate definition you might add some qualifier: a generally accepted pattern. But did you know that a pattern as simple as the Big Dipper is also known both as the Plow and as someone's wagon? That's not by civilizations far removed from one another, but by American and British viewers.

If a Socrates heckled you into an even more precise definition, you'd sooner or later add "as seen from Earth." The stars are not glued to the dome of the sky; some are relatively near, others far away. So the stars in any one constellation may have no more to do with one another than the moon and the tree over which it rises. Move a few steps to one side, and the moon rises over a different tree. Move around the universe, and the stars would group themselves into different patterns.

Even that definition—or any other you can think up—is unsatisfactory. It also includes asterisms.

You have already met some asterisms. The three stars in the middle of Orion, its belt, are considered an asterism. So is the Square of Pegasus. Three of the corners belong to the constellation Pegasus, the fourth—strangely—is borrowed from Andromeda (an-DROM-uh-duh).

Asterisms and constellations were invented as memory aids to make star finding easier.

They haven't done too well.

They are hard to learn. The moon, a street lamp, even the sky glow from the next town blot out the dimmer stars that

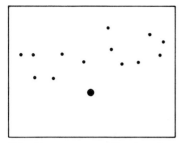

5.1 *Virgo—a group of fourteen stars.*

5.2 *Virgo—as a baroque constellation.*

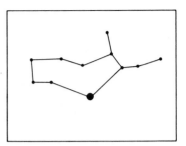

5.3 *Virgo—geometric pattern resembles snail.*

make up most constellations. Most of them don't look like the things after which they are named.

Here's an example: The Little Dog, Canis Minor (KAY-nis MEYE-nuhr), sports all of two stars. What kind of dog is that? A dachshund? A late-blooming kindergartener would draw it better. How can such a "pattern" help you find it again?

For many centuries constellations were drawn as fancy figures spread over the stars. Often the stars had little to do with the artist's subject. One star might mark the knee of a hero, another his elbow, but most of the stars were sprinkled over folds in his clothes. Not very helpful.

You can see why a book intended to teach the stars to future sea captains would avoid such fancy dress. It replaced baroque art with geometric designs. Most don't give a clue to the

5.4 *Virgo—stick figure by H.A. Rey.*

5.5 *Virgo—astronomer's boundaries.*

5.6 *Virgo—finding it from brightest star, Spica.*

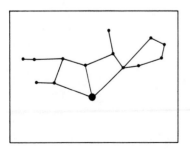

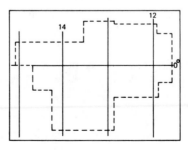

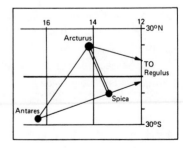

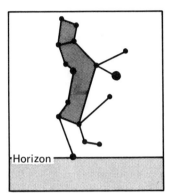
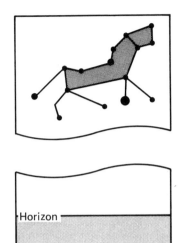
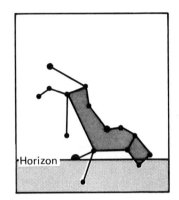

5.7   Tilt of constellations. Example: Leo the Lion as drawn by Mr. Rey. Leo rising (left) stands on its tail. Leo highest (middle) lopes. Leo setting (right) pounces.

name of the constellation. They are no easier to memorize than the star patterns themselves.

Then in 1952 Mr. H. A. Rey, with great imagination, redrew the lines to make the geometric figures look like human twins, a lion, a flying swan . . . .

With all constellation outlines—even Mr. Rey's helpful figures—you still have a problem. You have to allow for the angle of the figure changing as the constellation rises, bears more or less south, and sets (Figure 5.7).

It may surprise you that modern astronomers still use constellations. They use not the patterns of stars but boundaries settled by an international conference. The areas are drawn along hour lines and parallels of declination, resulting in rectangles with many jogs. So today every astronomer knows what area to sweep with his telescope if, for example, a comet is recovered in Canis Minor.

To find conspicuous constellations, this book uses the first-magnitude stars that you already know (Figure 5.8). I introduce the constellations in the order in which you met their prominent stars in the last chapter. Many of the star names are translated from Ptolemy's star catalog (circa AD 140), which has come to us through Arab astronomers.

OVERLEAF
5.8   Finding fourteen constellations from the fifteen brightest stars visible everywhere in the coverage area. All constellations—Mr. Rey's outlines— are shown bearing south and slightly enlarged from the strip map. The vertical 10° declination marks and the distance of the RA lines ($1^h = 15°$) let you measure distances in and between constellations. (One finger equals 2°; one hand 10°.) The small numerals label navigational stars (see Chapter 9.)

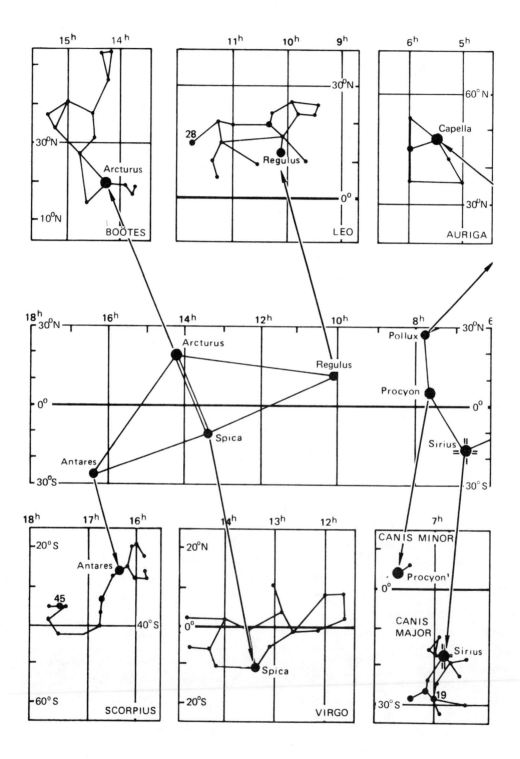

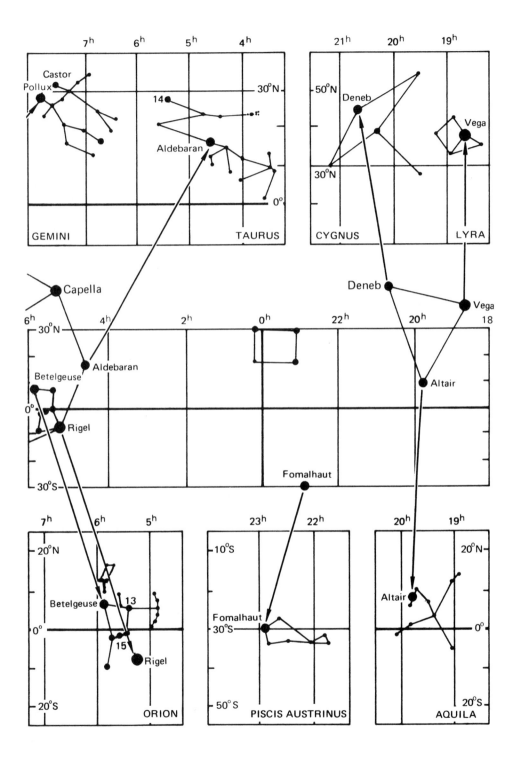

The Hexagon

Betelgeuse, the star near the center of the hexagon, is in the constellation Orion. In Arabic its name means "the shoulder of the central one." That's where Mr. Rey has drawn it.

Capella, meaning "little she-goat" in Latin, is in the constellation Auriga (or-EE-guh), the Charioteer. Did he use goats to draw his chariot? The image of a little she-goat pulling a chariot may help you connect the star name to the constellation.

Aldebaran in Arabic means "the follower." In the constellation Taurus (TO-ruhs), the Bull, Aldebaran follows the Pleiades or Seven Sisters. That asterism of dim stars is just hinted at near the 4^h line in Figure 5.8. You'll meet it properly in Chapter 10.

Rigel is in the constellation Orion, as is Betelgeuse. That explains why we can find only fourteen constellations from the fifteen first-magnitude stars visible everywhere in the area.

Rigel is Arabic for "foot." It's Orion's left foot (to your right), diagonally opposite Betelgeuse. Perhaps bottom right will remind you where in Orion to look for it. We'll meet another Arabic foot later in this chapter.

Sirius, the brightest fixed star, sometimes called the Dog Star, is in Canis Major (KAY-nis-MAY-jer), the Big Dog. One school has the word Sirius come from the same root as our verb "to sear." Try this memory aid: "Searing days, dog days, sirius, dog star in the Big Dog."

Procyon is in Canis Minor (KAY-nis MEYE-nuhr), by far the brighter of the two stars in that constellation. The name in Greek means "before the dog." In Greece Procyon rises before the Dog Star. Sirius rises farther south and later, like the sun in November.

Pollux is in Gemini (JEM-i-neye or JEM-in-nee), the Twins. There's no question about the word Pollux. It's from Greek mythology, the son of Leda and Zeus and twin of Castor. Most people say "Castor and Pollux," naming Castor first.

Here the twin on the left, Pollux, is the brighter of the two. It wasn't always so. One of the few recorded changes in brightness has made Castor dimmer than its twin.

The Double Triangle

Regulus is in the constellation Leo (LEE-oh), the Lion. In Latin Regulus means "little king," from "rex," a full-size king. To aid your memory you may think of the lion, king of beasts.

Spica is the brightest star in Virgo (VUHR-goh), the Virgin. Its Latin name means "spike," or "spike of grain" such as an ear of corn. The connection between spike and virgin is unclear. As a memory aid try "sacrificing a spike of corn in place of a virgin."

Antares leads you to Scorpius (SKOR-pi-uhs), the Scorpion. There's no question about this star's name. It's from the Greek, meaning rival of Aries, whom we know by the Latin name Mars. The rivalry between this star and the planet Mars is in their redness.

Scorpius is one of the few constellations that resembles its namesake. It needs little imagination to see in its stars a scorpion complete with stinger at the tail. Antares, once known as the Heart of the Scorpion, is in the head or near it.

For star watchers in the southern part of the area, Scorpius is a helpful constellation. It is almost always visible there when Orion is below the horizon. Already in latitude 35°N it rises within one hour of Orion's setting and sets about two hours before Orion rises. The periods when neither constellation is above the horizon get shorter as you go farther south.

In that part of the coverage area, if you just recognize these two constellations you'll have a friend in the sky most of the time. Either constellation lets you quickly align a strip map of the stars—printed or remembered—without bothering with local star time.

Arcturus is in Boötes (boh-OH-teez), the Bear Driver, also called the Herdsman. Arcturus—Latinized Greek—means "bear driver." You may recognize the root of "ursa." In Mr. Rey's pictures Boötes faces the Great Bear; perhaps you can remember the herdsman driving off the bear.

The Triangle

Vega is in the constellation Lyra (LEYE-ruh), the Lyre. The translation from the Arabic is no help: "the falling eagle" or "vulture." How does that get into a harp? Was Lyra once called the falling eagle to distinguish it from the next constellation?

Altair is in Aquila (AK-wi-luh), the Eagle. Altair translates "flying eagle" or "vulture." Memory aid: "The eagle flies at great altitude."

Deneb is in Cygnus (SIG-nuhs), the Swan. The Arabic word translates into "tail-of-the-hen," so the swan has a hen's tail. Don't let the drawing of the constellation mislead you: Deneb is in the tail. The long line is the bird's outstretched neck. That's how swans fly; just like their cousins, the geese. Cygnus aims to pass between Lyra and Aquila.

Some people find another asterism in Cygnus: the Northern Cross. Deneb and the stars of the neck form the long arm, the stars at the ends of the wings the short one.

On a star map or a list of stars you may come across a star named Denebola (de-NEB-oh-luh). That's not a variation of Deneb. Its Arabic for "tail of the lion." That's where Denebola is, in the tassel of the lion's tail, the star that's numbered 28 in Figure 5.8.

Fomalhaut

The star left out of the superconstellations, Fomalhaut, is in Piscis Austrinus (PISS-is Oss-TREE-nuhs), the Southern Fish.

Without instruments you need Fomalhaut to find that fish. It is off by itself, and its next brightest stars are of fourth magnitude. You already know how to locate Fomalhaut. It's in line with the western edge of the Square of Pegasus, and almost three spans south from its southwest corner.

This fish is spelled Piscis; the dim constellation of the zodiac is Pisces, the Fishes (plural, see Chapter 9).

Southern Stars

Here's a quick look at the constellations that contain the six southernmost first-magnitude stars.

Achernar is in Eridanus (e-RID-uh-nus), the River. Its name, from the Arabic, translates into "the mouth of the river." That's where it is, at the south end of the river Eridanus. Ptolemy called it simply "the River." Eridanus was a Greek river god, and an ancient name for the River Po. An old astronomy book I consulted calls the river Eridu.

The constellation Eridanus has also been called the rivers Jordan, Nile, and Styx. But it's always a river. That's all anyone can make of that pattern of stars. It meanders from almost the celestial equator near Rigel to about three spans south of it.

Except for Achernar there isn't a star brighter than third magnitude in the entire river bed. To fit this constellation, the astronomers' boundary of Eridanus makes no fewer than twenty right-angle jogs. That's what gives constellations a bad name.

Canopus, the second brightest star, almost on the same hour circle as Sirius, is in the constellation Carina (car-EE-nuh), the Keel. Canopus is the Greek name of an ancient city in northern Egypt, Kanopos. The Canopic branch of the Nile long ago silted up. The town died, leaving canopic jars for archeologists.

Carina is a modern constellation. Until 1930 it was part of Ptolemy's constellation Argo (ARR-goh), the Ship (of Jason and the Argonauts of Greek mythology). That year astronomers decided it was too big and broke it up like a mere telephone company. Even in the southernmost part of the area you won't see much more of the Ship's keel than Canopus.

Acrux here forms the foot of the long arm of Crux (KRUHKS), the Cross, often referred to as the Southern Cross. Acrux is a modern name—a cable address for this star, which hasn't had a common name for some time.

Astronomers referred to it as alpha in Crux. The Greek letter designation goes back to 1603 when Johann Bayer invented that scheme for naming stars. The general rule is this: The brightest star in a constellation gets the first letter of the Greek alphabet, alpha. The second brightest gets the second letter, beta, and so forth.

That makes Acrux alpha. But today's astronomers use the Latin possessive case for the constellations. To them Acrux is alpha Crucis. Examples of other Latin genitives are alpha Eridani for Achernar, alpha Carinae for Canopus, alpha Canis Minoris for Sirius.

That may not seem like a great brain saver, but it's surely easier to say beta Scorpii than "the northernmost star in Scorpion's forehead" as Bowditch described it.

You probably don't want to bother with the various Latin genitives. Why not call that star "beta in Scorpius"?

This explains how the brightest star in Crux became Acrux. Have you guessed the name of the second brightest one? Becrux. It wasn't really necessary to invent a name for it. It had a rather pretty name already, Mimosa.

But we now have Becrux, and Gacrux for the star at the head of the long arm of the cross. These names are used by navigators. You won't hear often about Delcrux, here at the right end of the short arm of the cross.

The constellation Crux disappoints many people who see it for the first time. It's about three fingers by two, about a third the size of Orion. The stars vary greatly in brightness. Alpha and beta are first magnitude, gamma second, and delta third. These four stars are about all there is to Crux.

Hadar and Rigil Kent (Kent is short for Kentauri), only about 2½ fingers apart, are in Centaurus (sen-TOR-uhs), the Centaur. Rigil, despite the spelling, is again an Arabic foot. That star is often referred to as Alpha Centauri, even by people who shun most other Bayer names. Hadar, or Beta Centauri the less bright one of the two, is the other forefoot, nearer the Cross.

The six first-magnitude stars visible only in the southern part of the area are in four constellations, with Crux and Centaurus each having two.

Achernar is little help in tracing the course of Eridanus, the River. Canopus is in Carina, the Keel, but in this area the constellation itself never rises fully. If you can see Acrux where you are, you see the whole constellation of Crux, the Cross. Alpha and Beta Centauri put a base under the many lesser stars west of Scorpius that make up the body of the Centaur.

First-Magnitude Stars and Their Constellations

Use the top part of the table to find the name of a constellation from a known star. Use the bottom part of the table for finding the names of the brightest stars from known constellations.

Star name	Constellation (Latin)	Constellation (English)
Achernar	Eridanus	River
Acrux	Crux	Southern Cross
Aldebaran	Taurus	Bull
Altair	Aquila	Eagle
Antares	Scorpius	Scorpion
Arcturus	Boötes	Bear Driver
Becrux (Mimosa)	Crux	Southern Cross
Betelgeuse	Orion	Orion
Canopus	Carina	Keel
Capella	Auriga	Charioteer
Deneb	Cygnus	Swan
Fomalhaut	Piscis Austrinus	Southern Fish
Hadar	Centaurus	Centaur
Pollux	Gemini	Twins
Procyon	Canis Minor	Little Dog
Regulus	Leo	Lion
Rigel	Orion	Orion
Rigil Kent	Centaurus	Centaur
Sirius	Canis Major	Big Dog
Spica	Virgo	Virgin
Vega	Lyra	Lyre

Constellation (Latin)	Constellation (English)	First-MAG Star(s)
Aquila	Eagle	Altair
Auriga	Charioteer	Capella
Boötes	Bear Driver	Arcturus
Canis Major	Large Dog	Sirius
Canis Minor	Little Dog	Procyon
Carina	Keel	Canopus
Centaurus	Centaur	Rigil Kent, Hadar
Crux	Southern Cross	Acrux, Becrux (Mimosa)
Cygnus	Swan	Deneb
Eridanus	River	Achernar
Gemini	Twins	Pollux
Leo	Lion	Regulus
Lyra	Lyre	Vega
Orion	Orion	Betelgeuse, Rigel
Piscis Austrinus	Southern Fish	Fomalhaut
Scorpius	Scorpion	Antares
Taurus	Bull	Aldebaran
Virgo	Virgin	Spica

If you watch the night sky close to the southern border of the area, you'll add at least two more constellations—Crux and Centaur—to the fourteen found before from first-magnitude stars. You'll meet a few more constellations in the next chapter.

Summary

- Constellations and asterisms are patterns we read into the stars, as seen from Earth, to make it easier to orient ourselves in the night sky.

- All constellations tilt as they move through the sky. Many are hard to identify when the moon or manmade lights blot out their dimmer stars.

- At your location, constellations well south of the celestial equator may never completely rise. The brightness of their stars will be dimmed when they are low in the sky.

- The fifteen first-magnitude stars that are visible anywhere in the area lead to fourteen constellations (Figure 5.8). Orion has two first-magnitude stars.

- The six first-magnitude stars visible only to observers in the southern part of the area are found in four constellations. Crux and Centaurus have two first-magnitude stars each.

6

The Northernmost Constellations

We started our study of the night sky by wrapping a strip map with the brightest stars over the celestial equator (Figure 3.2). I then filled in the gap left near your southern horizon (Figure 4.3). That's where another six first-magnitude stars may appear when you watch from the most southerly part of the area.

Even with those and three northern first-magnitude stars—Capella, Vega, and Deneb—I have left one part of the sky uncovered.

No first-magnitude stars are found there. No planets ever appear in that space. But in it are some good friends of every sky watcher in the northern hemisphere: the Polestar and the Big Dipper. Let's fill that empty space!

Call it the polar cap. On the star globe (Figure 1.2), it's the area that surrounds the upper end of the shaft around which the sphere turns.

In Figure 6.1 the polar cap resembles an open umbrella, with the hour circles for ribs. The tilt of its handle—like that of the shaft of the globe—depends on your latitude. But we don't want a globelike outside view because we are always under the sloping umbrella, looking up at its inside.

If you have a camera whose shutter you can keep open, you can take a time exposure of the stars (Figure 6.2).

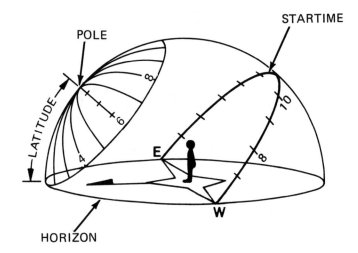

6.1   The polar cap of the sky. Star time is 12ʰ. The celestial equator, our reference line, is behind the observer looking north.

In such a photograph the stars will leave circular tracks. The length of the tracks will depend on how long the shutter was open. Stars farther from the center will leave longer tracks than stars nearer the center.

One star, very near the center, will seem to have hardly moved. That star—you guessed it—is the Polestar, Polaris (poh-LAYR-uhs), also called the North Star. It is very close to that important reference point in the sky—the celestial North Pole—around which all stars seem to revolve. The celestial North Pole is directly above the geographic North Pole on Earth.

Earth turning around its axis once every 23 hours and 56 minutes causes the apparent motion of all the stars. The motion is counterclockwise. That's a long word, but simpler than saying this: "In the upper part of the circle around Polaris the stars move from east to west, in the lower part from west to east."

Three facts help find the North Pole in the sky:

1. It is exactly north of you. (If you use a compass, you may have to correct for its declination at your location.)
2. It is as many degrees above your horizon as you are north of the equator on Earth. In latitude 40°N it will be two spans above the north point on your horizon.

3. Polaris is less than half a finger's width from the celestial North Pole.

That works in reverse, too: Find Polaris and you have located the celestial North Pole.

Sailors find their latitude by measuring how high Polaris is above the horizon. You can do the same. With your spans, hands, and fingers you can get your latitude. Subtract that from 90° and you have the height of the celestial equator above your south point. South is directly opposite Polaris.

Polaris is a second-magnitude star that remains clearly visible on a full-moon night. If you know approximate north, look for Polaris as many degrees as your approximate latitude above the north horizon. Great accuracy is not needed in either direction or latitude. There's no star bright enough to be confused with Polaris within more than one hand and three fingers from it.

There's another aid for finding Polaris.

Most people in the northern hemisphere recognize the Big Dipper (Figure 6.2). That, by the way, is not a constellation but an asterism. It's part of the constellation Ursa Major (UHR-suh-MAY-jer), the Big Bear.

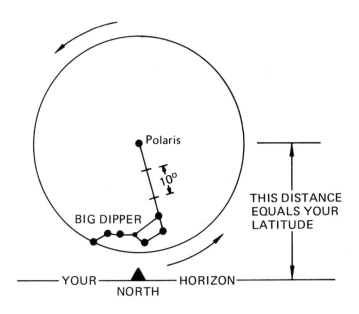

6.2 *Finding Polaris, which is within a half finger of the celestial North Pole. The pole is due North and as many degrees above the horizon as your latitude is north of the equator. (Five hands in the drawing show the view at 50°N latitude.)*

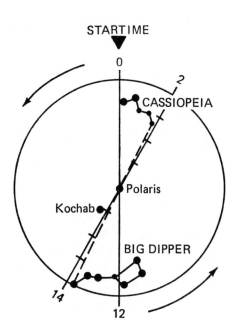

6.3 *Polaris, Kochab, Cassiopeia, and the Big Dipper—their relative positions and direction of rotation. Polaris and Kochab stay above the horizon at all times in the entire area. In the southern part of the area, Cassiopeia will be high in the sky when the Big Dipper disappears below the horizon.*

The two stars that form the side of the dipper that's away from the handle almost point at Polaris. Hence their name: Pointers. They are about 5½°—less than three fingers—apart. Polaris is five times that distance—not quite three hands—from the Pointer nearest it.

The Dipper will not always be below Polaris and "holding water" as in Figure 6.2 at star time 0ʰ. Twelve hours later, if it were still dark, you'd see it "spilling water" on Polaris below it.

About eight fingers from Polaris in the direction of the hole in the handle is a star about as bright as Polaris. Its name is Kochab (KOH-kab), "the north shining star" in Arabic. Kochab, Polaris, and a few dimmer stars form another dipper-shaped pattern, Ursa Minor (UHR-suh MEYE-nuhr), the Little Bear. Its shape is hard to make out unless conditions are good.

On the other side of Polaris, away from Kochab and the Little Dipper, is a much brighter constellation: Cassiopeia (Kass-i-oh-PEE-uh). Five of its stars, about as bright as Polaris, are supposed to look like that ancient queen sitting in a chair. You'll see the chair. You may also see it as a sloppy letter M when it's above Polaris; below Polaris you'd read it as a lazy W.

North of about latitude 40°N both Cassiopeia and the Big Dipper will be above the horizon at all hours of every night. The learned way to say that is "they become circumpolar" at about that latitude.

There you can find, or confirm your finding, of Polaris. It is almost on a line from the end of the handle (four hands away) to the stars at the dim end of the W or M (three hands away). (See Figure 6.4.)

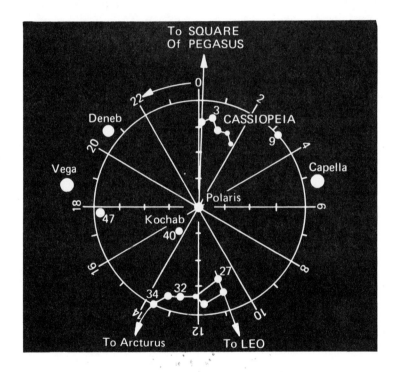

6.4 *The northernmost stars. Within four hands of Polaris all second-magnitude stars, and a few dimmer ones, are shown. The small numerals refer to navigational stars. (See Chapter 9.)*

The three first-magnitude stars north of the strip map are shown outside the circle. Arrows lead you to Arcturus, the center of Leo, and the eastern edge of the Square of Pegasus.

The height of the celestial pole above your horizon depends on your latitude. It will sink lower and lower as you move south. There must be a point where the Big Dipper will begin to sink below the horizon when it is directly below Polaris. The star at the end of the handle, being the farthest from the pole, will be the first to disappear below the horizon. Other stars will follow it. Near latitude 28°N, say near Tampa, Florida, the whole Dipper dips out of sight at times.

Summary

• This chapter fills the void around the celestial North Pole.

• The polar cap lacks first-magnitude stars. Sun, moon, and planets never get near it.

• Polaris is so close to the pole that all stars seem to revolve around it.

• Its nearness to the pole also lets you find North and your latitude from Polaris.

• The Pointers in the Big Dipper make finding Polaris easy. When they are not visible the constellation Cassiopeia helps.

7

Magnitude of Stars and Planets

Up to now we have looked at fixed stars, bodies that make their own light like the sun. From far away our sun would look like a star; close by, a star would look like the sun. But the nearest stars are several hundred thousand times farther from us than the sun.

The stars keep station almost as if they were attached to the dome of the sky. An astronomer of ancient Persia, Egypt, or China would instantly recognize familiar patterns, his constellations, and asterisms in tonight's sky.

The ancient astronomer would have to look around the sky to find the planets. He probably didn't know that planets, like Earth, get their light from the sun. He certainly didn't call them planets, a name the Greeks invented that means "wanderers." The planets seem to roam among the fixed stars.

If you saw a very bright star that's not on your map, you'd be looking at a planet. But that's not an efficient way to find planets. And you wouldn't know which planet you were looking at.

You may have heard of a simple way to tell a planet from a star: The fixed stars twinkle, the planets don't. Unfortunately that test doesn't always work. When the ground is warm and the air is very cold, even the planets twinkle in the rising air currents. It's an effect similar to the one that makes a landscape shimmer when seen above a hot asphalt road.

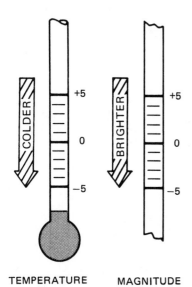

TEMPERATURE   MAGNITUDE

*7.1   Minus values. Left thermometer
scale shows plus and minus degrees;
right shows the corresponding magnitude scale. Colder than zero is minus
degrees; brighter than zero is minus
magnitude.*

Seen from a boat in the tropics, with water and night air at
almost the same temperature, the fixed stars have no cause to
twinkle. For the same reason, for many hours of many nights
the fixed stars shine with a steady light over land, say in Florida.

There are only five planets visible to the naked eye.

Their brightness often helps distinguish them from the fixed
stars and from one another, so let's look in detail at the astronomer's measure of brightness: magnitude.

In earlier chapters you became familiar with the brightest—
first magnitude—stars, the ones that make up the superconstellations, for example. You also met a number of less bright—
second magnitude—stars, in the Big Dipper and Cassiopeia, for
example. You also saw brighter stars shown more prominently
on star maps than dimmer ones.

Modern astronomers list the magnitude of stars to at least
one decimal, for example 2.1 for Polaris. Any stars between
magnitudes 1.5 and 2.5—for example, Polaris—are classed as
second magnitude.

A star between 2.5 and 3.5 is classed as third magnitude. The
faintest stars visible under good conditions, without optical aids,
are of fifth magnitude (4.5 to 5.5).

The scale doesn't end there. The world's largest telescopes
can bring in stars of twenty-second magnitude (21.5 to 22.5).

Nor does the scale stop at the bright end.

By custom, which goes back to the earliest surviving star
catalog, stars brighter than second magnitude are classed as first-
magnitude stars. That takes in quite a range. A star just a little
brighter than Castor (1.6) is a first-magnitude star; for example,
Spica (1.2), or Altair (0.9). Very much brighter Vega (0.l) is still
considered first magnitude.

What about even brighter stars? The scale now continues like
the one on a thermometer. When it gets colder than zero
degrees (Fahrenheit or Celsius) you use minus figures. After zero
magnitude come minus magnitudes (Figure 7.1).

Only two fixed stars rate a minus number: Canopus (– 0.9),
visible only in the southern part of the area and Sirius (– 1.6).
Both are still called first-magnitude stars. That classification includes stars from – 1.6 down to + 1.5.

By the way, when there can be no doubt, you can leave out

the plus signs, as I did until I came to Canopus and Sirius. Just as complaining about the heat you don't have to say or write it was plus 100° Fahrenheit or plus 40° Celsius in the shade.

At first, minus numbers—you'll meet many more when you read about the magnitude of planets—seem needlessly confusing. Why not start the scale at the brightest planet, when it's brightest, to get rid of minus magnitudes? That would make first magnitude Procyon a fifth-magnitude star. Even more confusing!

Other scales have been proposed. But every star atlas, star globe, star finder, or report of the discovery of a comet uses the magnitude scale I've sketched in Figure 7.2.

As you know, magnitude—despite the word's origin—has nothing whatever to do with size. It's a measure of brightness of a celestial body as seen from Earth. (The "as seen" in "as seen from Earth" means seen by the human eye, as opposed to measurements on a photographic plate or by some electronic device.) Antares (1.2) has about twice the diameter of Arcturus (0.2) but is a whole magnitude dimmer. Jupiter's diameter is about ten times that of Venus, but its magnitude is about two units less.

More precisely, it's the magnitude you would assign when a star or planet was high in the sky. Theoretically, you should measure it with the body directly overhead where it appears brightest. Practically, from about halfway up in the sky the difference between observed magnitude and zenith magnitude is negligible.

But close to the horizon all stars and planets seem dimmed—quite apart from haze. Their light near the horizon has a longer path through the absorbing atmosphere of Earth. That's also what dims and reddens the rising and setting sun.

A fixed star or planet 10° above the horizon loses one magnitude. A second-magnitude star appears as a third-magnitude star. A star or planet 6° above the horizon loses 1½ magnitudes; 4° from the horizon it loses two magnitudes.

The fixed stars we have met have—with some exceptions—the same magnitude year in and year out. Betelgeuse is such an exception; it varies from 0.4 down to 1.3 with an average period of 5.7 years.

The same stars had—again with a few exceptions—the same magnitude in Ptolemy's day. You have already read about Castor slipping from first magnitude to just below that class.

The planets, on the other hand, change their magnitude noticeably within a few months. You'll read about the reasons for these changes, and their rate of repetition, in the next chapter. Here we are only interested in the range of magnitudes, a planet's greatest and least brightness.

Looking at the right side of Figure 7.2, near the bottom, you'll see that the magnitude of Venus varies from − 4.4 at brightest to − 3.3 at dimmest. That makes Venus by far the brightest starlike object in the sky. It's enough to identify Venus. It will be the first "star" to appear in the evening, or the last to fade in the morning.

Jupiter at its brightest is of magnitude − 2.5, at its dimmest − 1.2. That makes it most of the time brighter than the brightest star, Sirius (− 1.6). And always brighter than the second brightest star, Canopus (− 0.9). It never gets within several hands of either star, so its brightness alone may serve to identify Jupiter.

Mars has by far the greatest range of magnitude of all the planets. Occasionally it outshines Jupiter at its brightest; at other times it's only about as bright as Polaris. But at all times Mars has a distinctly reddish tint.

You may not see much difference in the color of stars. Many people don't. But most people see the redness of Mars.

A good test for your color vision at night is the constellation Orion: Betelgeuse (at the top) is on the red side, Rigel (the foot) on the blue. When Orion is below the horizon, there's another test: Antares is as red as Mars, Spica and Altair are both bluish.

Saturn, the last of the easily seen planets, is always as bright as a first-magnitude star, − 0.4 at most, + 1.4 at least.

Mercury here will only rarely apear bright enough to attract your attention. When it's visible to the naked eye it varies in magnitude from about − 1.4 down to 0.7. That seems fairly bright. But Mercury will always be close to the horizon—in the

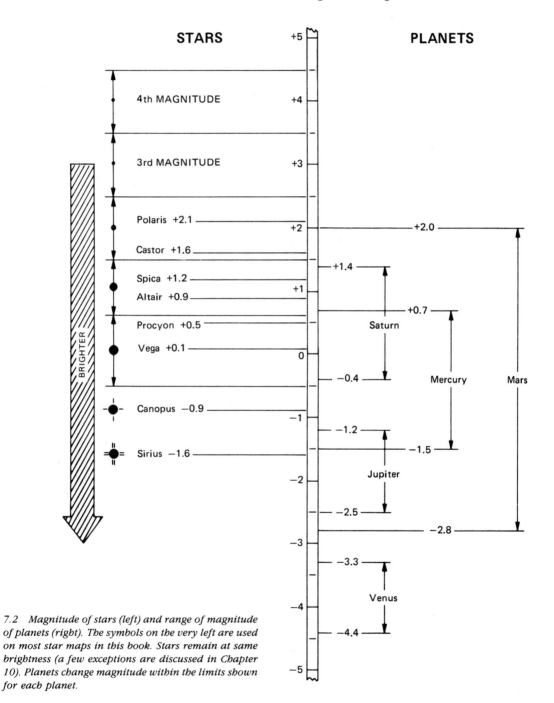

7.2 *Magnitude of stars (left) and range of magnitude of planets (right). The symbols on the very left are used on most star maps in this book. Stars remain at same brightness (a few exceptions are discussed in Chapter 10). Planets change magnitude within the limits shown for each planet.*

Planet Identification from Magnitude and Color

Rules	Notes
1. A *very bright* star seen in sky area covered by the strip map, properly placed, may not be a fixed star, it may be a planet.	The placement of the strip map is shown in Figure 3.2; its bright stars in Figure 4.1. Planets are never seen outside the area of the strip map.
2. If it is *not Sirius,* it must be a planet.	Sirius in the Hexagon, south and east of Orion, is near 7ʰ, S 16°.
3. If it is *extremely bright* it is Venus.	Check: Venus can never be North or South. It may bear about East for up to three hours before sunrise, or bear about West for up to three hours after sunset.
4. If it seems *reddish,* it is Mars.	Don't confuse it with Antares, (16.5ʰ, S 26°) in December 1991, January 1992, and the late fall of 1993, 1995, 1997, and 1999.
5. If its light is *not reddish* it can be either Jupiter or Saturn.	Jupiter is always brighter. Saturn is East of Jupiter until June 2000 when Jupiter overtakes.
6. If it is brighter than Sirius it is Jupiter.	Jupiter until 2000 is without doubt brighter at all times.
7. If it's not Jupiter it's Saturn.	

eastern sky before sunrise or the western sky after sunset. That will make it seem less bright.

How magnitude can help you identify the planets is shown in the above table. When and where to find all five naked-eye planets is the subject of the next chapter.

Summary

• Magnitude is a measure of the brightness of a star, or planet, as it appears to the eye on Earth. The lower the number, the brighter the object.

- In order of increasing brightness, after magnitude $+2$ comes $+1$, then 0, then -1, -2, and so on.

- Stars of magnitude $+5$ are just visible to the naked eye when conditions are favorable.

- The brightest fixed star, Sirius, has a magnitude of -1.6; it is one of two fixed stars with minus magnitude. Canopus (-0.9) is the other.

- The magnitude of most fixed stars remains constant.

- The magnitude of the planets visible to the naked eye, though variable, helps to identify them.

- Stars and planets near the horizon appear dimmer than when they are high in the sky.

8

The Planets

Most people are more interested in the planets than in fixed stars. I can prove that. When I was director of a planetarium, visitors and telephone callers would ask me, "What's the star I see now in the west about . . . ?" At least nine times in ten it turned out to be not a star but a planet.

That's understandable: The brightest objects draw the most attention. There are other reasons, I think.

There are only five planets to be seen in the sky with the naked eye, but an awful lot of fixed stars. (That awful lot isn't as large as most people estimate; not even an uncountable number. Under the best conditions we see no more than a couple thousand stars at any time.)

Stars are very distant fiery balls of gas—not easy to get personally excited about. Planets are near, and they are brothers and sisters of our own Earth. Unmanned space probes have flown past, measured, and photographed the naked-eye planets, even landed on Mars. That gets the planets in the news. A manned space flight to Mars is technically possible now, a trip to even the nearest star is still science fiction.

Also, people who look at a fixed star through a telescope are usually disappointed; it's just another pinpoint of light. The planets—even in a small telescope—are something else. You see the sickle shape of Venus, the polar snowfields of Mars changing with the seasons, the flattening of Jupiter and its mysterious red spot, the rings of Saturn . . . .

There are nine major planets. Astronomers also keep track of more than one thousand minor planets—most of them between the orbits of Mars and Jupiter. They range in size from a few hundred miles down to the limits of visibility.

Here's a sentence to remind you of the first letter of the names of the major planets:

> *My Very Educated Mother*
> *Just Showed Us Nine Planets*

That gives you—in the order of increasing distance from the sun—Mercury, Venus, Earth, Mars, Jupiter, Saturn, Uranus, Neptune, and Pluto.

I don't know who made up this memory aid, but I know that mother indeed was very educated. She picked one of the rare times when all nine planets were visible. She must also have had access to a powerful telescope. How else could she have found Pluto (magnitude 14)?

Here I'll talk only about the planets you can see without a telescope. You already know the five—Mercury, Venus, Mars, Jupiter, and Saturn—that you can see wandering against the background of the fixed stars. The wandering has a simple explanation: The planets—including our own planet Earth—revolve around the sun (see Figure 8.1).

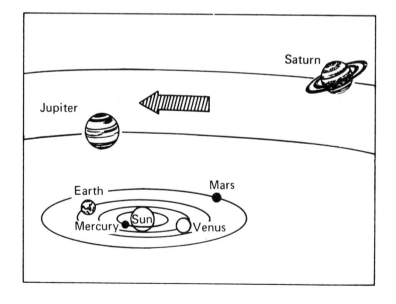

8.1 *Six planets and their orbits around the sun. Besides Earth, that includes all the planets visible to the naked eye. All revolve in the same direction (arrow) and in about the same plane.*

In the drawing they have been much enlarged—all to the same scale—with respect to the sun. The diameter of Earth is less than 1/100th that of the sun.

You may credit Copernicus with first publishing that explanation in 1543. It had been proposed already by Aristarch of Samos (310–230 BC). In 1609 and 1619 Kepler gave us formulas that fit the observed orbits of the planets.

In 1687 Newton explained the cause of their movements: gravity.

Miles, or kilometers, are a poor measure for the distances we observe in the sky; the figures get too large.

Already for our nearest neighbor in space, the moon, which orbits Earth as the planets orbit the sun, the figure is about one quarter million statute miles (four hundred thousand kilometers).

A radio signal travels at the speed of light—186,000 statute miles (300,000 km) in one second. It takes a little more than one second to reach the moon from Earth, or to reach Earth from the moon. That's why you barely noticed the delay between Houston Space Center's questions and the answers from the astronauts on the moon.

The same signal would take about three minutes to reach us from the nearest planet, Venus, when it's closest to Earth.

From the nearest fixed star the light takes more than four years to reach us. That's about one hundred million times as long as from the moon. And that's from the nearest star!

Finding or Identifying Planets

To identify a bright planet that you see where no bright star is shown on the strip map, you don't need to know any of that. And you don't need to know anything about the laws of motion. But you may want to know in advance where among the stars you can find the planets tonight.

You are like the astronomer who needs to know where to point his telescope or radio dish to observe a particular planet. He looks it up in an annual reference book. He doesn't care how the book's been calculated from the Kepler–Newton equations with minor refinements added.

He looks for the same data we have used in locating stars in the sky from our star map:

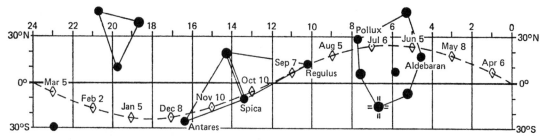

8.2   The ecliptic—the apparent annual path of the sun among the stars.

- right ascension, the local star time when the planet will be due south
- declination, the planet's distance north or south of the celestial equator

For observation of the stars—by telescope or with the naked eye—it's simplest to imagine the Earth standing still while the dome of the sky turns. In the same way, it's simplest to consider the Earth standing still while the sun and the planets move around it.

Figure 8.2 traces the sun's path among the stars as seen from Earth. Of course, we don't see the stars and the sun at the same time. But from the stars that disappear before sunrise and the ones that appear after sunset we can estimate the position of the sun for any day of the year. It'll be almost the same next year on that day, and the year after.

At the beginning of spring, about March 20, the sun is at 0^h RA, in declination 0°. That's no coincidence. The beginning of spring is defined as the moment when the sun, northbound, crosses the celestial equator, declination 0°. The 0^h circle, also by definition, is at that crossing point: the vernal equinox, the first point of Aries, or simply Aries.

For the next six months the path of the sun will be north of the celestial equator. It reaches its highest point—declination about 23½°N—at RA 6^h around June 21. That marks the beginning of summer.

At the beginning of fall, about September 22, the sun now southbound crosses the celestial equator at RA 12^h.

For the next six months the path of the sun will be south of the celestial equator. It reaches its lowest point—declination about 23½°S—at RA 18^h. That marks the beginning of winter about December 21.

Figure 8.2 explains why, say, Regulus is not visible on August 23, or Antares on November 30. The sun on these dates seems right near the places of these two stars. For about ten days before and after these days you won't see these stars.

You say you could do without that information. Nobody needs a map to find the sun. And nobody will look for the stars—or planets—until it's dark. Why then bring the path of the sun into this discussion of the planets?

Simply this:

The planets will always be seen close to the
line traced by the apparent annual motion
of the sun, called the ecliptic.

That explains an earlier statement: The planets are always within the area of our strip map, never north near Deneb, Vega, or Capella, nor in the polar cap, nor in the gap south of the map near the southern horizon.

If the orbits of the naked-eye planets were exactly in the Earth–sun plane, they would follow the ecliptic exactly. As it is, all of them except Mercury will be seen within a finger or so from the ecliptic.

That brings these planets at times close to the first-magnitude stars near the ecliptic: Aldebaran and Pollux in the Hexagon; Regulus, Spica, and Antares at the base of the Double Triangle.

The five stars near the ecliptic by coincidence are all of about the same magnitude (1.2). The calendar in Appendix B gives you the magnitudes of the planets for many years into the future. That often lets you tell Mars, Jupiter, and Saturn from one of the nearby reference stars.

In October 1998, for example, Mars at then magnitude 1.7 will get quite close to Regulus. Since 1.7 is half a magnitude dimmer than 1.2, Regulus should be clearly the brighter of the two objects.

There's another point in the sky that sometimes helps, the

point where the northbound sun crosses the celestial equator at RA 0ʰ, which is also 24ʰ.

The beginning of this book tells how to trace the equator in the sky. The eastern (left) edge of the Square of Pegasus lies almost on the 0ʰ circle. So that edge points straight down to the vernal equinox on the equator (see Figure 4.2).

In the next few pages I'll show you how to find the position and magnitude of the five naked-eye planets. Each of them has its own character. It would be logical for me to take them up in their order from the sun—Mercury, Venus, Mars, Jupiter, Saturn. Instead I'll start with the planet that comes closest to Earth, Venus. That makes talking about Mercury easier.

Venus

Venus, with a magnitude that varies between − 3.3 and − 4.4, is brighter than any other planet or fixed star. That's why at times it is the first "star" to appear in the evening, at others it's the last to disappear in the morning. That's also why it's called the evening star and the morning star.

At the time of its greatest brightness, you can see Venus in daylight.

One way to do that is to follow it in the morning. It's clearly visible in the brightest dawn. Don't take your eyes off it; you'll still be able to see it when the sun rises. Keep looking at it and you can follow it as long as your patience lasts.

Another method works whether Venus is morning or evening star. Take a cardboard tube, such as calendars are mailed in, and search the sky where Venus should be. To avoid eye damage, don't look at the sun directly.

If you see Venus at sunrise, you'll have a fair idea where to look for it later. It will be to the right and higher than the sun at the same distance from it as at sunrise. It will be about the same distance from the sun for several days.

If you have seen Venus after sunset yesterday, look to the left and higher than the sun. It'll be about the same distance from the sun as it was last night. Again it will be about the same distance from the sun for several days.

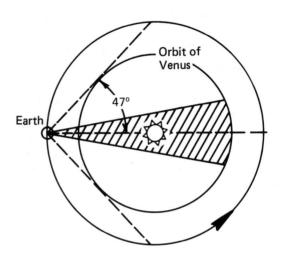

8.3 Orbits of Venus and Earth seen from above. The arrow on the orbit of Earth shows the direction in which both planets move around the sun. But you can imagine the Earth standing still—at the small circle—and only Venus moving. In each circuit around the sun, Venus shows two periods when it is too close to the sun (in the shaded area) to be seen.

Seen from Earth, Venus can never get more than 47° from the sun. That makes it visible to the naked eye for only a few hours after sunset or before sunrise.

When Venus is in the area shown at the top part of the drawing, it will set (and rise) after the sun and will be visible after sunset in the western sky.

When Venus is in the area shown in the bottom part of the drawing, it will rise (and set) before the sun and will be visible before sunrise in the eastern sky.

The orbits of Venus and Earth are sketched in Figure 8.3. Their orbits and speeds of revolution around the sun combine to make Venus evening star for several months. It will then disappear in the glare of the sun for a couple of weeks.

If the orbit of Venus were exactly in the Earth–sun plane, in the middle of that period Venus would pass over the face of the sun. Actually such a transit of Venus is a rare event. The last one was in 1882. Until 2004, Venus will always pass close above or below the disk of the sun.

After Venus emerges from the glare of the sun it becomes morning star for several months. Then it disappears for several months, passing at the far side of the sun. Then the cycle, which has taken about nineteen months, repeats.

When Venus becomes invisible on the far side of the sun it is about six times as far from us then as when it disappears on the near side. (You can check that in Figure 8.3.) You might expect that to cause its change of magnitude, but it is more complicated than that.

When Venus is near us (left in the sketch) you expect it to be very bright. But its sunlit side is almost turned away from us. All we see is a slim crescent like the moon just before or after new moon. When Venus is close to its maximum distance from us (right in the sketch) you expect it not to be very bright.

But we see the disk almost fully lit by the sun, like an almost full moon.

These phases—visible in even a small telescope—counteract the change in brightness due to distance. The result is that Venus is brightest about five weeks before and after it gets between us and the sun.

To find Venus on any particular day you only need to know whether to look for it in the morning or evening. Appendix B gives that information and the magnitude at mid-month through the year 2000. Here is the key to these entries:

AM   Venus is visible before sunrise in the east
PM   Venus is visible after sunset in the west
-o-   Venus is too close to the sun for observation
to   Venus disappears in sun's glare about this date
after   Venus emerges from sun's glare about this date

For example, AM to 26 – 3.9 tells that Venus is morning star until the 26th of this month. The – 3.9 means magnitude at mid-month. PM after 27 – 4.2 tells that Venus is evening star after the 27th, and on that day is of magnitude – 4.2.

After the last year given in the calendar you can still get the needed information by using data for eight years earlier. For September 2001, for example, use the data for September 1993.

Mercury

Mercury is as hard to find as Venus is easy.

Copernicus, who lived to be seventy, never managed to see it. You may want to give up, skip Mercury, and go on to Mars.

Mercury's greatest magnitude is given in some books as – 1.8. That is misleading—it's not visible at that time. Even – 1.4 is optimistic. Mercury then is on the border of naked-eye visibility near the sun.

Venus, as you just read, can never get more than 47° from the sun, as seen from Earth. For Mercury, closer to the sun, the limit is 28°. Even that only happens at seven-year intervals.

The orbits of Earth and of Venus are almost circles. The orbit of Mercury is noticeably elliptical. That makes its maximum distance from the sun, as seen from Earth, change from one round trip to the next. That distance—greatest elongation is the astronomer's term for it—can be as little as 18°.

When stars become visible in the evening, or disappear in the morning, the sun is 6° below the horizon. Does that mean that Mercury when its elongation is 18° is then 12° above the horizon? No. The line from Mercury to the sun slopes; that brings Mercury closer to the horizon.

In the course of one year there are six or seven greatest elongations of Mercury. That gives about half-a-dozen periods during which you can see Mercury. Half these periods will be before sunrise, half after sunset. Until Mercury is about 10° away from the sun you can't see it at all. That leaves from a few days to a few weeks for each of these periods of visibility.

At best you'll see Mercury for about one-half hour low above the eastern horizon before dawn, or low above the western horizon after dusk. You'll recall that the light of stars and planets is dimmed by one or two whole magnitudes when a body is seen between 10° and 4° from the horizon. That makes Mercury—during much of its periods of visibility—look no brighter than a second-magnitude star.

Even when you know the periods of visibility and whether to look for Mercury in the morning or in the evening, you can miss seeing it. Copernicus, who knew the data, was clouded out in Prague. So was I, time and again, in the much more favorable Bahamas. There overcast skies are rare but clouds low on the horizon are common.

Some readers will take that as a challenge.

The favorable periods are not easy to find. One well-known source is the *Nautical Almanac,* published annually jointly by the U.S. Naval Observatory and the Royal Greenwich Observatory. I could find Mercury only during part of each period given there. Navigators, by the way, don't use Mercury, and these almanacs give no daily predictions for it.

Appendix C gives the approximate starting and ending dates of each morning and evening period, and the magnitude of Mercury on these dates.

In the North Temperate Zone the before-sunrise periods are most favorable for observation when they fall in early October, the after-sunset periods when they fall in early April.

After the last part in the table you can still get approximate periods of visibility and magnitude of Mercury:

> *Take the data for thirteen years earlier,*
> *then add two days to the dates given.*

For example: The favorable 1990, March 29 to April 15 evening period will be repeated in 2003, between March 31 and April 17.

Mars

At times Mars is hard to overlook. At other times it's hard to find.

That's caused by great shifts in its brightness. Its magnitude can vary from − 2.8 (brighter than Jupiter at its brightest) down to + 2.0 (about equal to Polaris).

Once you have found it, its reddish cast will confirm that you have the right object. Among the fixed stars only Antares (magnitude 1.2) is that red. Since Antares is one of the bright stars near the ecliptic, Antares and Mars occasionally meet in the sky. That gives you an opportunity to compare their color. You may agree that Antares is well named: the rival of Mars.

In Figure 8.1 you see that the orbit of Mars is outside the orbit of Earth. So the geometry that keeps the inner planets, seen from Earth, within so many degrees from the sun, does not apply to Mars.

That means Mars can be visible not only near the rising or setting sun but at any time of night.

Mars is only invisible when it's close to the sun—in the shaded area of Figure 8.4. The motions of Earth and Mars combine for that to happen every other year and last several months.

For example, Mars is too near the sun for observation in 1991 from the beginning of October to early December; in 1993 from late November to early February 1994; and in 1995 from mid-January to the end of April. For such periods Appendix B uses the symbol "-o-" and "to" and "after" dates as for Venus.

When Mars emerges from the glare of the sun, it is far from us. The distance of Earth to sun adds to the distance of Sun to Mars. That's when Mars will appear dim.

Then the distance diminishes, Mars becomes brighter. The fact that less of the lighted side of Mars is turned toward us doesn't matter much. Unlike Venus, Mars never appears half lighted or crescent shaped. In a telescope it looks at worst like the moon a couple of days before or after full moon.

Mars will continue to brighten until it is opposite the sun, fully lighted, and nearest Earth. At that time Mars, 180° from the sun, will bear south twelve hours after the sun, at midnight. The time from one such opposition to the next is two years and a few weeks.

The orbit of Mars is quite elliptical. So its distance from both the sun and from us—and with it its magnitude—will differ from one opposition to the next. The brightest oppositions are about sixteen years apart. The last spectacular one was in September 1988 (magnitude − 2.5).

Here are the next oppositions with their magnitudes:

November 1990	− 1.8
January 1993	− 1.2
February 1995	− 1.0
March 1997	− 1.0
April 1999	− 1.4

In one month Mars moves against the background of the stars on the average about one hour in right ascension (1½ hands) eastward.

But strange things happen when Mars is near opposition. If you could look down on the solar system, you'd see Earth moving faster on the inner track, overtaking slower Mars on the outer track.

Seen from Earth, Mars seems to slow down, stand still, then move backwards to the west. It then seems to slow again and stand still again before taking up its normal eastward journey amid the stars.

The calendar in Appendix B lists the positions of Mars and its magnitude for the 5th, 15th, and 25th of every month. You

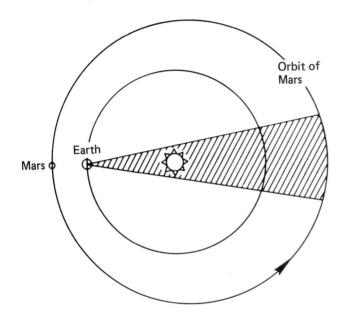

Orbit of
Mars

Earth

Mars

8.4 *Orbits of Mars and Earth seen from above. The arrow on the orbit of Mars shows the direction in which both planets travel. But you can imagine the Earth standing still—at the small circle—and only Mars moving.*

In each circuit around the sun, Mars will show one period when it is too close to the sun (in the shaded area) to be seen. Mars can be seen from Earth at any hour of darkness. Its orbit is noticeably elliptical, unlike those of Earth and Venus, which are nearly circular.

can see an example of that retrograde motion by looking at its RA for the months of October 1990 to March 1991.

The positions of the five reference stars near the ecliptic (see Figure 8.2) and footnoted in Appendix B will often help you find the position of Mars. In the example, the zigzagging of Mars takes place about three fingers north of Aldebaran.

When no reference star seems near the position of Mars, you may try to pencil in its position on Figure 8.2. Mars will always be within a finger or two of the ecliptic. So just marking the proper RA on the ecliptic will do. Often you'll discover that it's not so very far from one of the reference stars or a star in a constellation you know already.

Mars is hard to find when you are looking for a second-magnitude star with a red cast. Make it easier on yourself: Look for it when it's bright. One month later it will be about a hand and a half east of where you found it.

Here are the periods when Mars is of first magnitude:

March 1990 through March 1991

March 1992 through April 1993

January 1994 through July 1995

December 1995 through April 1996

November 1996 through April 1998
December 1998 through February 2000

There is no easy way for finding the position and magnitude of Mars after our calendar runs out.

Jupiter

Jupiter (magnitude -1.2 to -2.5) is almost always the brightest starlike object after Venus. Like Mars it can be seen at any angle from the sun. About every thirteen months Jupiter reaches opposition. It is then at its brightest, is visible all night, and bears south at midnight.

Jupiter is visible near the ecliptic for about eleven months of the year. Then for about one month it gets too close to the sun for observation. Except just before and after these disappearances Jupiter is brighter than the brightest fixed star, Sirius (-1.6).

Year	Jupiter Nearest Month	MAG	Not Visible Month
1989	December	-2.4	June
1990	————		July
1991	January	-2.1	August
1992	February	-2.0	September
1993	March	-2.0	October
1994	April	-2.0	November
1995	June	-2.1	December
1996	July	-2.2	————
1997	August	-2.4	January
1998	September	-2.5	Feb 10–Mar 10
1999	October	-2.5	Mar 18–Apr 15
2000	November	-2.4	Apr 24–May 20

Through the telescope Jupiter always looks virtually full. Its change in magnitude is almost entirely due to its changing

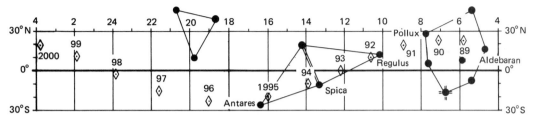

8.5 Jupiter's midyear position among the stars (1989–2000). The calendar in Appendix B gives monthly positions and magnitudes.

distance from us. When Jupiter is closest—at opposition—that distance is about four times our distance from the sun; when Jupiter is farthest, it is about six times that distance.

It takes Jupiter almost twelve years to complete a circle tour of the sky. So in one year it changes right ascension only about two hours (three hands) toward the east. Near opposition it stalls, moves westward, and stalls again before resuming its eastward trek, just like Mars.

The calendar in Appendix B gives the position and magnitude of Jupiter as for Mars, but only for the 15th of each month.

The slow movement of Jupiter lets you keep track of it easily. After a couple of weeks or a month you'll find it just about where you saw it last.

If you haven't seen Jupiter lately, Figure 8.5 will help.

It charts the year-to-year position of Jupiter among the stars. Until July of any year it will be a little right of the plotted position (toward last year's position). Later it will be a little left of the midyear mark (toward next year's midyear mark).

When you run off that map, or after Appendix B expires at the end of the year 2000, use the data for twelve years earlier. For 2003, for example, use 1991 for the midyear position, also for the month of greatest brightness (January), and for Jupiter's absence from the night sky (August).

Saturn

Most of the time Saturn looks like a bright first-magnitude fixed star.

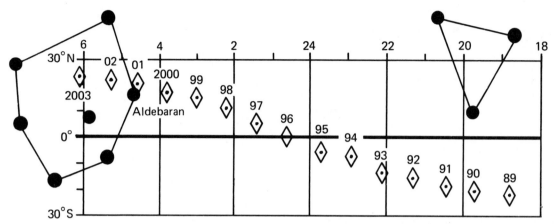

8.6 *Saturn's midyear position among the stars (1989–2000). Appendix B gives monthly positions and magnitudes.*

It's never dazzling like Venus—or even as bright as Jupiter at its dimmest—and (unlike Mars) shows no particular color. Despite all that, you can find or identify Saturn without much effort.

• In 1990 and 1991 Saturn is two or three hands below Altair. The superconstellation Triangle seems to point directly at it (see Figure 8.6). In 1994 through 1996 Saturn is less than two hands below the Square of Pegasus. In 1998 it is about a span west of Aldebaran, one of the two reference stars in the Hexagon. Saturn keeps approaching Aldebaran and catches up with it in mid-2001.

• The very slow movement of Saturn helps find it. It takes about thirty years to return to the same place amid the stars. So in one year it moves only a little more than one hand.

  For example, when you have nailed down the position of Saturn in relation to a star pattern in June, you'll find it quite near there in December, and only a little farther east next June.

• Appendix B gives Saturn's midmonth position and magnitude, just as for Jupiter.

• From 1990 to 2000 Saturn is brighter than any nearby fixed stars virtually all the time.

At the beginning of the period it is already about three hands east of the reference star Antares. It doesn't get near the next ecliptic reference star, Aldebaran, until the end of the century.

The only first-magnitude star near its path is Fomalhaut (1.3). In the springs of 1994, 1995, and 1996 Fomalhaut and Saturn may be hard to tell apart by magnitude alone. But Saturn is always the one nearer to the Square of Pegasus.

• Jupiter enters the west area of Figure 8.6 at the beginning of 1996 when Saturn is better than halfway across it. Jupiter, by far the brighter, keeps gaining but stays west of Saturn until the end of May 2000. Then Jupiter, a finger north, overtakes.

Saturn is nearest Earth about every 12½ months. See the table below for dates and the planet's magnitude at these oppositions. It also gives the periods when Saturn is hidden from us in the glare of the sun.

Year	Saturn Nearest Date	MAG	Not Visible Date
1990	Jul 14	+ 0.3	Jan 1–Jan 18
1991	Jul 27	0.3	Jan 7–Jan 31
1992	Aug 7	0.4	Jan 19–Feb 11
1993	Aug 19	0.5	Jan 30–Feb 21
1994	Sep 1	0.7	Feb 10–Mar 5
1995	Sep 14	0.9	Feb 23–Mar 18
1996	Sep 26	0.7	Mar 7–Mar 29
1997	Oct 10	0.4	Mar 20–Apr 11
1998	Oct 23	0.2	Apr 2–Apr 25
1999	Nov 6	+ 0.0	Apr 15–May 10
2000	Nov 19	− 0.2	Apr 29–May 23

Near the time of opposition Saturn, like Mars and Jupiter, halts then runs backward and halts again before getting back on its eastward course.

If you are used to Saturn's staying near a place for a long time, you may not notice that movement. But you can verify it in Appendix B. In 1995 in mid-June its RA is 23.7, in July it's

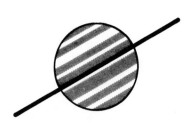

8.7 *Saturn's rings seen from Earth at maximum opening (top) and edge on (bottom). Since the rings are brighter than the planet itself, the magnitude of Saturn depends much on the opening of the rings, which change from maximum to minimum in about seven years.*

still 23.7, in August 23.6, in September 23.5, and so on to November's 23.3. It's the same in December. The RA then slowly increases to get back in February to where it had been in August.

Predicting the magnitude of Saturn is complicated business. Only a small part of Saturn's range in brightness is explained by the planet's changing distance from the sun and from us. The greater part of change in magnitude—Saturn always appears full to us—is due to its rings.

Every 14¾ years—give or take a year—the rings are edge on to observers on Earth. The rings then contribute nothing to the total magnitude of the planet. That last happened in 1980.

On May 22, 1995, the Earth passes through the ring plane southbound. On August 11 it crosses the other way, and on February 12, 1996, it again crosses it southbound. Sometimes the rings completely disappear from view in a telescope; in other years they just leave a knife edge of light. For several months in that year Saturn will be a bit dimmer than the comparison stars near the ecliptic.

The rings then open again, and already at the opposition in November 2000 you'll see Saturn as magnitude – 0.2, almost as bright as it ever gets.

There's no quick way to calculate the position and magnitude of Saturn after the calendar in Appendix B expires.

Summary

- Only five planets are visible to the naked eye.

- You won't find Mercury unless you look for it, and perhaps not even then. It can only be seen during periods lasting from a few days to a few weeks. It will then be low in the eastern sky before dawn, or low in the western sky after dusk.

- Venus is unmistakably the brightest starlike object. It's visible for several months in the eastern sky for up to three hours before sunrise, or—again for several months—in the western sky for up to three hours after sunset.

- Mars, Jupiter, and Saturn are always near the ecliptic, the sun's apparent path among the stars (Figure 8.2).

- Mars, always reddish, is at times very bright.

- Jupiter is almost always brighter than the brightest fixed star, but dimmer than Venus.

- Saturn, always of first magnitude, stays in the same area of the sky for many months.

- The calendar in Appendix B helps you find or identify the planets through the year 2000.

9

Dimmer Stars and Constellations

Having worked your way through the book this far, you can now find or recognize all the planets visible to the naked eye, all the first-magnitude stars, and about a score of constellations.

That puts you into a select class of experts, ahead of many professional navigators.

Do you want to know more? What, for instance, about the constellations of the zodiac? Their names, in Latin or English or both, are familiar to millions who couldn't tell Venus from Vega.

"Zodiac" means circle of animals (from the same root as zoo). There are five very strange animals in that zoo: twins, a virgin, a pair of scales, an archer, and a water carrier.

When the Babylonians started all this some 4,000 years ago, at the beginning of spring the sun crossed the celestial equator in the constellation Aries. That's why we still start the list with Aries.

For telling where along the ecliptic a planet is on a given day, these constellations are of little use. The constellation Virgo, for example, stretches three times as far as the constellation Cancer.

So the ancients divided that area near the ecliptic more precisely into twelve "signs," each 30° wide. Quite logically they named each after one of the constellations.

At least since Hipparchus, who lived in the second century

Latin	Pronunciation	English
Aries	AR-eez or AR-ee-ez	Ram
Taurus	TO-ruhs	Bull
Gemini	JEM-i-neye or -nee	Twins
Cancer	KAN-suhr	Crab
Leo	LEE-oh	Lion
Virgo	VUHR-goh	Virgin
Libra	LEE-bruh	Scales
Scorpius	SKOR-pi-uhs	Scorpion
Sagittarius	saj-i-TAY-ree-uhs	Archer
Capricornus	kap-ree-KOR-nuhs	Sea Goat
Aquarius	uh-KWAYR-ee-uhs	Water Carrier
Pisces	PEYE-seez	Fishes

BC, western astronomers have known about the precession of the vernal equinox. At the beginning of spring the sun crosses the equator a little farther west than the year before. (After about 25,800 years it returns to the same point on the equator.)

More than 2,000 years ago that crossing point—still called the First Point of Aries, or Aries for short—moved out of Aries into Pisces. It soon will slip out of Pisces into Aquarius.

If your birthday is April 1, the sun was in the constellation Pisces when you were born. But by your sign you are still an Aries. When the sun comes to be in Aquarius, people born on that day will still be called Aries.

You can find five constellations of the zodiac—Taurus, Gemini, Leo, Virgo, and Scorpius—from the first-magnitude stars they contain. That's the five stars—all of magnitude 1.2— we used in Chapter 8 to help find the planets.

The other constellations of the zodiac range from hard to find to very hard to find. Only two of them contain even second-magnitude stars (Figure 9.1).

Sagittarius sports two second-magnitude stars: Kaus Australis (KOS os-TRAY-lis), #48, less than a hand east of the bright star in the tail of the Scorpion; and Nunki (NUHN-kee), #50, about a hand northwest of Kaus Australis (Figure 9.2).

Aries has one second-magnitude star: Hamal (HAM-uhl), #6,

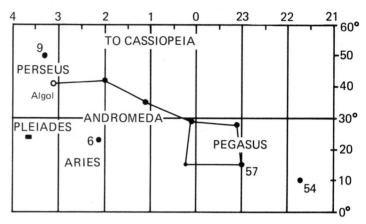

9.1   Second-magnitude stars near the Square of Pegasus. Some of them form another dipper, shown here enlarged from the scale used in the other star maps. Numbers refer to navigational stars.

meaning "lamb" in Arabic, two hands east of the Square of Pegasus.

The Square of Pegasus also can help you find the dim constellation Aquarius. It straddles the celestial equator between the Square and Fomalhaut.

The remaining four constellations of the zodiac, are best found from the others.

Cancer, impossible to find from its own stars, lies in the space between Gemini and Leo, two easily found constellations.

Libra, according to its place on the list, must be between Virgo and Scorpius. Its brightest star, #39, has a long name: Zubenelgenubi (ZOO-BEN-el-je-NOO-bee). That's Arabic, meaning "the southern claw of the scorpion." Perhaps that'll remind you that it's near Scorpius—but it's west and somewhat north, not south.

Capricornus is east of Sagittarius and hard to find. Its northwest corner, near which are two of its three brightest (third-magnitude) stars, is two hands south of Altair and a bit east. To the east it borders on Aquarius, which you have found from the Square of Pegasus and Fomalhaut.

Pisces, between Aquarius on the west and Aries on the east, straddles the equator.

Finding the Constellations of the Zodiac

Aries	Two hours behind Square of Pegasus, #6
Taurus	Aldebaran, first-magnitude star
Gemini	Pollux, first-magnitude star
Cancer	Between Gemini and Leo
Leo	Regulus, first-magnitude star
Virgo	Spica, first-magnitude star
Libra	Between Virgo and Scorpius
Scorpius	Antares, first-magnitude star
Sagittarius	Scorpius' tail, one hand west, #48, #50
Capricornus	South of Altair, east of Sagittarius
Aquarius	Between Square of Pegasus and Fomalhaut
Pisces	Between Aquarius and Aries

You may think me arbitrary for emphasizing first-magnitude stars and seemingly slighting stars of second magnitude and fainter. After all who can unfailingly tell the difference? I certainly can't.

But I've not been quite as arbitrary as it seems. Here and there I have slipped in second-magnitude stars. Very early I introduced Castor the twin of Pollux, stars in the constellation Orion, and the asterism Square of Pegasus (Figure 4.2). In the constellation outlines (Figure 5.8) you met many more less prominent stars. And all the stars in the polar cap (Figure 6.4) were second or third magnitude.

There are a few dimmer stars in the vicinity of the Square of Pegasus (Figure 9.1). All told you have already met more than half of all the second-magnitude stars.

Figure 9.2 shows all the second-magnitude stars within the strip map and immediately north of it. All the second-magnitude stars north of that are shown in Figure 6.4.

What about such stars south of the strip map? They too are all shown. In most of this area only a few will ever appear bright enough for you to think of them as second-magnitude stars. The best time—and for some of these stars the only time—to see them is when they are highest in the sky, when they are about south of you.

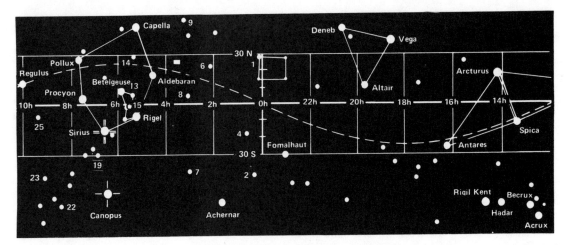

9.2 *All second-magnitude stars within the strip map and near its south and north borders. The northernmost second-magnitude stars are all shown in Figure 6.4. Numbers refer to navigational stars.*

Quite a few are in what used to be the constellation Argo (AR-goh), the Ship. Now they are in the constellations created by its breakup. They include Puppis (PUHP-is), the Stern; Vela (VEE-luh), the Sails; Carina (kar-EE-nuh), the Keel . . . .

Canopus, the second brightest fixed star (magnitude − 0.9), is in Carina. It is easily seen in central and southern Florida, highest when local star time is about 6½ʰ.

The Navigational Stars

Your interest in the stars may be for their use in celestial navigation. That's the method for getting your location—on land or in the air but mostly at sea. Measure the height of known stars above your horizon and note the exact time of each observation. (See Chapter 17.)

Do you know enough stars for that? The answer is pleasant: You already know more stars than you'll ever need.

A navigator will shoot stars for which he can find the exact position in the daily pages of the *Nautical Almanac* or in the *Air Almanac*. Both list the 57 navigational stars, which include

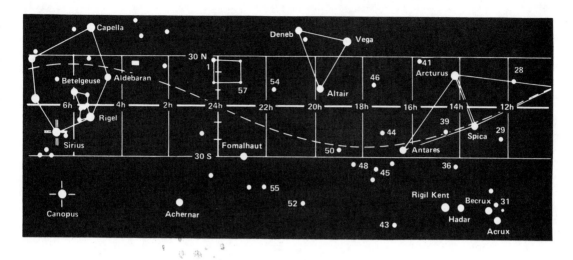

of course the first-magnitude stars you already know well. (A fine point: Instead of Becrux they include Gacrux, which is just over the border into second magnitude.)

You'll find most of the less bright navigational stars and their numbers in Figure 9.2. The remaining ones are shown in Figure 6.4. They are listed in the following tables.

You'll note in both drawings that the numbers increase with increasing right ascension. That's most noticeable in the Square of Pegasus. The star nearest RA 0^h is #1; the one near RA 23^h is #57. That helps find these stars on the star maps, and sometimes in the sky too.

Figures 6.4, 9.1 and 9.2 show all the second-magnitude stars ever visible in the North Temperate Zone and well beyond it, both toward the North Pole and the equator. You'll find the remaining southern stars, visible only farther south, in Figure 14.3.

The navigational stars include a few third-magnitude stars. Markab, #57, the southwest corner of the Square of Pegasus, is an example.

You won't find Polaris among the navigational stars. Not that navigators don't use it. On the contrary. It's great for finding

Navigational Stars (in numerical order)*

#	Star	MAG	Pronunciation	Constellation
1	Alpheratz	2.2	al-FEE-ratz	Andromeda
2	Ankaa	2.4	AN-kuh	Phoenix
3	Schedar	2.5	SHED-uhr	Cassiopeia
4	Diphda	2.2	DIF-duh	Cetus
6	Hamal	2.2	HAM-uhl	Aries
7	Acamar	3.1	AY-kuh-mar	Eridanus
8	Menkar	2.8	MEN-kar	Cetus
9	Mirfak	1.9	MEER-fak	Perseus
13	Bellatrix	1.7	bel-LAY-tricks	Orion
14	Elnath	1.8	EL-nath	Taurus
15	Alnilam	1.8	AL-ni-lam	Orion
19	Adhara	1.6	uh-DAY-ruh	Canis Major
22	Avior	1.7	AY-vi-or	Carina
23	Suhail	2.2	soo-HAYL	Vela
24	Miaplacidus	1.8	MEYE-a-PLAS-i-duhs	Carina
25	Alphard	2.2	AL-fard	Hydra
27	Dubhe	2.0	DUHB-ee or DOO-bee	Ursa Major
28	Denebola	2.2	de-NEB-oh-luh	Leo
29	Gienah	2.8	JEE-nuh	Corvus
31	Gacrux	1.6	GAY-kruhks	Crux
32	Alioth	1.7	AL-i-oth	Ursa Major
34	Alkaid	1.9	al-KAYD	Ursa Major
36	Menkent	2.3	MEN-kent	Centaurus
39	Zubenelgenubi	2.9	ZOO-BEN-el-je-NOO-bee	Libra
40	Kochab	2.2	KOH-kab	Ursa Minor
41	Alphecca	2.3	al-FEK-uh	Corona Borealis
43	Atria	1.9	AT-ri-uh	Triangulum Australe
44	Sabik	2.6	SAY-bik	Ophiuchus
45	Shaula	1.7	SHO-luh	Scorpius
46	Rasalhague	2.1	RUHS-al-HAYG-wee	Ophiuchus
47	Eltanin	2.4	el-TAY-nin	Draco
48	Kaus Australis	2.0	KOS os-TRAY-lis	Sagittarius
50	Nunki	2.1	NUHN-kee	Sagittarius
52	Peacock	2.1	PEE-kok	Pavo
54	Enif	2.5	EN-if	Pegasus
55	Al Na'ir	2.2	al-NAYR	Grus
57	Markab	2.6	MAR-kab	Pegasus

For first-magnitude navigational stars see Appendix E

Navigational Stars (in alphabetical order)*

MAG	Star	#	Meaning
3.1	Acamar	7	The end of the river
1.6	Adhara	19	The virgins
1.7	Alioth	32	The fat tail of the eastern sheep
1.9	Alkaid	34	The leader
2.2	Al Na'ir	55	The lighted one
1.8	Alnilam	15	One of the string of pearls
2.2	Alphard	25	The solitary star of the serpent
2.3	Alphecca	41	The broken one
2.2	Alpheratz	1	The horse's navel
2.4	Ankaa	2	Mythological bird
1.9	Atria	43	Alpha in Triangulum Australe
1.7	Avior	22	Coined modern name
1.7	Bellatrix	13	Female warrior
2.2	Denebola	28	Tail of the lion
2.2	Diphda	4	Frog
2.0	Dubhe	27	She bear
1.8	Elnath	14	The one who butts
2.4	Eltanin	47	The dragon
2.5	Enif	54	The horse's nose
1.6	Gacrux	31	Gamma in Crux
2.8	Gienah	29	The raven's wing
2.2	Hamal	6	The lamb
2.0	Kaus Australis	48	Southern bow
2.2	Kochab	40	North shining star
2.6	Markab	57	Saddle
2.8	Menkar	8	The hen's beak
2.3	Menkent	36	Shoulder of the Centaur
1.8	Miaplacidus	24	The still water
1.9	Mirfak	9	The elbow of the Pleiades
2.1	Nunki	50	The shoulder
2.1	Peacock	52	Peacock
2.1	Rasalhague	46	The head of the snake charmer
2.6	Sabik	44	The conquerer
2.5	Schedar	3	The breast of the lady in the chair
1.7	Shaula	45	The stinger on the scorpion's tail
2.2	Suhail	23	Little plain
2.9	Zubenelgenubi	39	Southern claw of the scorpion

For first-magnitude navigational stars see Appendix E

Constellations from Navigational Stars

#	Star Name	MAG	Constellation (Latin)	Pronunciation	Constellation (English)
2	Ankaa	2.4	Phoenix	FEE-niks	Phoenix
4	Diphda	2.2	Cetus	SEE-tuhs	Whale
8	Menkar	2.8	Cetus		
9	Mirfak	1.9	Perseus	PUHR-ee-uhs or PUHR-soos	Perseus
23	Suhail	2.2	Vela	VEE-luh	Sails
25	Alphard	2.2	Hydra	HEYE-druh	Sea Serpent
29	Gienah	2.8	Corvus	KOR-vuhs	Crow
41	Alphecca	2.3	Corona Borealis	koh-ROH-nuh boh-ri-AL-is	Northern Crown
43	Atria	1.9	Triangulum Australe	trey-an-goo-luhm os-TRAY-lee	Southern Triangle
44	Sabik	2.6	Ophiuchus	off-i-OO-kuhs	Serpent Holder
46	Rasalhague	2.1	Ophiuchus		
47	Eltanin	2.4	Draco	DRAY-koh	Dragon
52	Peacock	2.1	Pavo	PAY-vo	Peacock
55	Al Na'ir	2.2	Grus	GRUHS	Crane

your latitude—just as we did with hands and fingers—and it lets surveyors find true north.

A three-page table in the *Nautical Almanac* gives the corrections for Polaris not being quite at the celestial North Pole. The method is simpler than for any other star.

If constellations interest you more than individual stars, see the Constellations from Navigational Stars table.

Summary

- Of the twelve constellations of the zodiac along the ecliptic—where the sun, the naked-eye planets, and the moon move—five are found from their first-magnitude stars. They are Taurus, Gemini, Leo, Virgo, and Scorpius. Two more of these constellations, Sagittarius and Aries, can be located from their second-magnitude stars. The other, dimmer, ones can be found from their neighbors.

- Figure 9.1 shows how to find a few more second-magnitude stars.

- For celestial navigation, most navigators use first-magnitude stars whenever possible.

- All 57 navigational stars visible in the North Temperate Zone and near it are mapped in Figures 9.2 and 6.4.

- The dimmer navigational stars let you locate more than another dozen constellations.

10

The Moon

Why have a chapter about the moon? Nobody needs a book to find or identify it.

Also, the moon doesn't fit the title of this book. It's neither a star nor a planet. I shall not weasel out of that by reminding you that the ancient astronomers had the moon—and the sun—classed with the other "wanderers" in the sky. If they spoke Greek, they called wanderers "planetai."

Then why have a chapter, however brief, about the moon? Simply this: When the moon lights up the sky, you can see only the brightest stars and very few constellations. You certainly can't enjoy the dimmer objects discussed in Chapter 11.

So it helps the practical star watcher to know beforehand when the moon will be above the horizon and how bright it'll be. That's why this chapter is here.

Everybody knows the moon goes through phases.

You can demonstrate the phases by taking a light-colored ball or apple to model the moon and a light source—or the sun itself—to model the sun. You play the part of an observer on Earth.

1. Face the sun. Hold the ball between thumb and middle finger of the left hand. Hold the ball, at eye level in front of you. The side of the ball facing you will not be lit by the sun: *New Moon.*

    If you held the ball directly in front of the sun you would have simulated an eclipse of the sun. In most months the

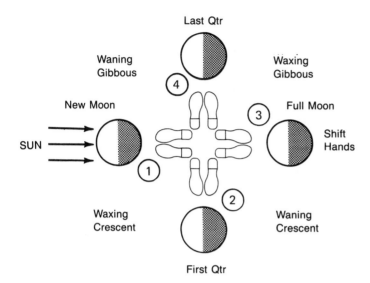

10.1 The four main phases of the moon, illustrated with a ball and a light source.

moon in its orbit, which forms an angle of 5° with the ecliptic, passes above or below the sun.

2. Turn 90° to the left. The ball is lit on the right half by the sun, the left half is unlit: *First Quarter Moon.*

    Don't let the word *quarter* confuse you. We call it quarter because we see it about one quarter into the cycle, which lasts roughly a month (29¼ to 29¾ days). First quarter then is about one week after new moon.

3. Turn another 90° to get the sun behind you. The ball is now fully lighted: *Full Moon* (roughly two weeks after new moon).

4. Shift hands, holding the ball between thumb and middle finger of the right hand. Turn another 90° left. The ball now is lighted on the left half, the right half is dark: *Last Quarter Moon* (roughly three weeks after new moon).

5. Another quarter turn gets you back to the next new moon, roughly one month after the last one.

To describe the moon between these main phases we need a few more terms. Two are fairly obvious. Waxing describes the lighted part of the moon increasing in size, as it does from

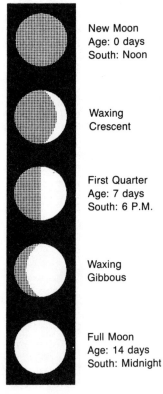

New Moon
Age: 0 days
South: Noon

Waxing
Crescent

First Quarter
Age: 7 days
South: 6 P.M.

Waxing
Gibbous

Full Moon
Age: 14 days
South: Midnight

10.2 Phases of the moon, with its approximate "age" (days since new moon) and time when it bears south.

new to full moon. Waning describes it shrinking in size, as it does between full and new moon.

The sickle shape of the moon in its first or last quarter is the crescent phase. If you are not familiar with sickles, perhaps you know croissants, fluffy pastries of similar shape. That's the French version of the same Latin word.

Gibbous (with the first letter sounded hard, as in give) describes a moon that appears more than half lighted, as in the week before and after full moon.

For star watching you don't care much about the phases of the moon as such. But you do want to know when the moon will be above the horizon. The two are closely linked.

Let's start with a new moon (age zero days). You won't see the moon then, except during an eclipse of the sun. It's too close to the sun. The moon will be south with the sun at noon. And it rises and sets with the sun.

You probably won't become aware of the moon until a couple of days later when it forms a thin crescent—unlighted on the left—east of the sun. It will set not long after the sun and will not spoil the viewing of stars.

The exact times of moonset and moonrise, like those of sunset and sunrise, depend on your latitude and how far north of the celestial equator the moon or the sun happens to be on that day.

On the average the moon bears south of you about one hour later each day. The exact average is 52 minutes. But that hardly matters since it varies a good deal from average—from about half an hour to about an hour and a quarter.

The moon will stay up longer in the western evening sky while it gets fatter. That's the waxing crescent stage.

When the moon is about a week old—at first quarter—with its left half still dark, it will be due south of you at 6 PM. It will be up and quite bright during the evening hours.

Then the moon, now in the waxing gibbous phase, will seem to grow, get brighter, and bear south an hour later each night. The moon will interfere with your finding dim objects during all but the last hours of the night.

About two weeks after new moon, the moon will be fully

lighted, opposite the sun. Having risen at sunset, it will bear south at midnight, and set at sunrise. The moon, now at its brightest, overwhelms all but the brightest stars. That spoils finding most constellations all night.

The full moon, eliminating the lesser stars, makes finding the superconstellations easy. When you spot an object about as bright or brighter than the leftover first-magnitude stars, you can be sure you have found one of the planets.

After having been full, the moon in its third week is west of the sun. In this waning gibbous phase it looks eaten away on the right side and a little more eaten away each night. It rises later each night and so lets you find dim objects in the first hours of darkness.

A week after full moon, at last quarter, the moon is lit on the left half, dark on the right. You'll have the whole first part of the night to study dim objects in the sky.

In the waning crescent phase that follows, the moon bears south in the morning hours. That adds night by night to the hours without moonlight.

Two weeks after full moon, four since the last new moon, the moon again will be on the same bearing as the sun. It will have disappeared in the glare of the morning sun a day or so before. After the new moon, the cycle starts all over.

Obviously a sliver of moon, say two days before or after new moon, gives considerably less light than a moon at first or last quarter. When the moon is full, it has twice as much surface illuminated as at either quarter. That makes it considerably brighter.

If you thought one full moon seemed much brighter than some others, you were right. It was not the company of the evening that made it so. The moon's orbit is elliptical. That makes the moon appear twenty percent brighter when full moon comes near perigee (least distance from Earth) than when it comes near apogee (greatest distance from Earth).

At times, especially when you should see only a small crescent moon, you see the dark part faintly lit. The poetic name for this is "the old moon in the arms of the new."

Astronomers call it earthshine. When the moon is a small

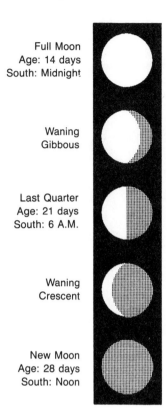

Full Moon
Age: 14 days
South: Midnight

Waning
Gibbous

Last Quarter
Age: 21 days
South: 6 A.M.

Waning
Crescent

New Moon
Age: 28 days
South: Noon

crescent—waxing or waning—an astronaut on the moon would see Earth almost full. The ashen light (astronomers' ghostly term, not mine) is sunlight reflected by Earth.

The calendar in Appendix B gives the dates of the four phases of the moon in the order of their appearance. For example:

Moon	Last 4	New 12
	First 20	Full 27

Last stands for last quarter; New for new moon; First for first quarter; and Full for full moon.

If a phase falls near the very beginning of a month, there will be the same phase again near the end. Say there's a full moon on July 1; there'll be another on or about July 30.

The dates are calculated for Universal Time (UT), about what we used to call Greenwich Mean Time. It's the time to use when you want to give dates regardless of the observer's longitude. The only drawback is this: Our calendar may at times differ by one day from a locally published calendar.

For example: Full moon is given for the 11th when the moon is in line with the sun at 3:00 AM UT. A calendar published in the United States may use Eastern Standard Time, five hours slow on Greenwich. That would make the full moon at 10:00 PM on the day before, the 10th. For star watching that makes no difference. The moon would look virtually full on either day—and spoil looking for dim objects all night.

Summary

• Here is help for finding dark-of-the-moon periods for looking for dim objects in the sky. I simplified the table by making each phase exactly seven days. It's easier to remember that way.

Phase of Moon	Age Days	Dark Period
New moon	0	All night
Waxing crescent		All but first hours
First quarter	7	Last half of night
Waxing gibbous		Last hours of night
Full moon	14	No time of night
Waning gibbous		First hours of night
Last quarter	21	First half of night
Waning crescent		All but last hours
New moon	28	All night

• Dates of moon phases are given in Appendix B for every month through the year 2000.

11

More Night Sky Objects

The Milky Way

The Milky Way can be seen in the dark of the moon as a shimmering river.

The Greeks called it the galactic circle—after "gala," milk. "Circle" describes it better than our "way"; it's a band all around the celestial sphere.

If you have a globe handy you can get an idea of its location. Imagine an airline route or use a string to get the shortest routes between points. It starts at Iceland, runs via Panama to a point near the Antarctic Circle south of New Zealand. From there it follows the shortest route to Sri Lanka (the former Ceylon), and back to Iceland.

The Milky Way follows a similar course (Figure 11.1). Astronomers speak of a galactic equator and a band about 20° (two hands) wide. It narrows and widens, is interrupted by dark islands, and over a long stretch splits into two arms.

The axis of this river comes closest (27°, about 2½ hands) to the North Pole near RA 0^h in Cassiopeia (Iceland on the globe). It runs between Gemini and Orion (Panama) toward the Southern Cross. There (corresponding to the point south of New Zealand) the axis of the Milky Way comes within 27° of the celestial South Pole. It then passes through Scorpius, Aquila (Sri Lanka), and Cygnus back to Cassiopeia (Iceland).

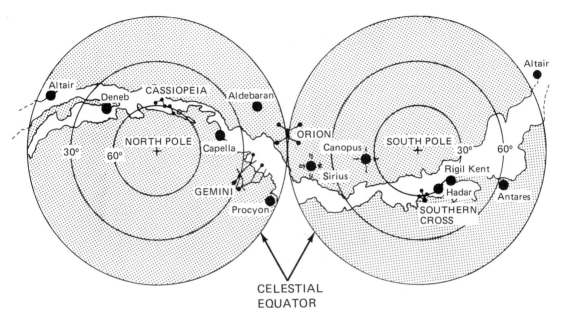

11.1   The Milky Way circles the celestial sphere. Left: the north half of the sphere. Right: the south half. The two halves are joined at the celestial equator, the outermost circle in both views.

All these constellations, with the exception of Cassiopeia, contain first-magnitude stars. In fact, sixteen of the twenty-one brightest stars lie in, or near, the Milky Way.

That could be coincidence but isn't.

The diameter of our island universe of stars, our galaxy, is about six times its height in the center. Chances of finding any star, including a first-magnitude star, are greatest in the equatorial plane.

Until Galileo in 1609 looked at it through his telescope, the nature of the shimmering band in the sky was a mystery. Today all you need is a sky free of manmade or moon light and a pair of binoculars to see what it is: stars and more stars too dim to be seen separately by the unaided eye. To make such a band that wide across the entire sky takes a lot of stars. Estimate: One hundred thousand million stars!

In a way looking for the Milky Way is more challenging than looking for planets. Planets always stay in our basic strip map.

The Milky Way winds about from the polar cap through the strip and into the southern gap.

That makes for unexpected variety from hour to hour, from month to month, from season to season. The hours when the moon is below the horizon add an element of rarity to the prey. Local lights make enjoyment of the Milky Way impossible, while you can still find the planets easily.

Then there is the weather. When the moon is dark and you are free, finally, for an evening or have set the alarm for an unusually early hour just to watch the sky . . . it clouds over.

There are, of course, times more favorable than others for watching the Milky Way.

In the cold season in this area, local star time 5^h is a good time to watch—morning or evening—if it's dark. The Milky Way that runs from your northwest horizon to between Orion and Gemini, and on to your southeast horizon.

In the warm season in this area local star time 20^h is an even more interesting time to watch—morning or evening—if it's dark. The Milky Way then runs from your northeast horizon through the single Triangle, where it splits in two. Then it continues to your southwest horizon with its widest area in Scorpius.

Other Galaxies

Ours is not the only galaxy. There are an estimated two thousand million others.

The two nearest the Large and the Small Magellanic Clouds—from which light takes "only" 170,000 years to reach us—can't be seen from this area.

But when conditions are good, you can see half a dozen external galaxies from anywhere in the North Temperate Zone.

Look, for instance, about three fingers northwest from the star shown near RA 1^h in Figure 9.2. You'll see a small hazy patch. That's the famous Great Nebula in Andromeda M31, and two companions. In modern terms it is not a nebula but—like ours—a spiral galaxy.

When you have found M31, you are looking at the most distant object visible to the naked eye, more than two million light years away. The light you see there tonight left that long ago!

Another—quite different—nebula is M42, the Great Nebula in Orion. From the belt of Orion, directly below its middle "star," hangs his sword. The third "star" down the sword looks too fuzzy to be a star.

We now know that it is a mass of gas made to glow by some very hot stars embedded in it. It is 1,600 light years from us and the brightest of the hundreds of fuzzy patches cataloged in our galaxy.

Some of the stars in the area of the Orion nebula are believed to have come into being as stars less than a million years ago. Earth is at least five thousand times older. And there's good evidence that inside the nebula stars are still being born right now.

An Open Cluster

The small black rectangle near 4^h in Figure 9.1 marks the Pleiades. Once they were considered a constellation, the Pleiades or the Seven Sisters, or in South America the Seven Little Goats. By whatever name, even downgraded as they are now to the rank of asterism, they make an interesting naked-eye object.

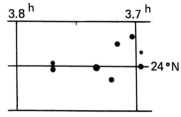

11.2 *The Pleiades, enlarged 50 times from our usual map scale, which would run them all together.*

Figure 11.2 shows how one can count seven stars. Don't worry if you can only see six or so. One sister or goatlet is said to have lost most of her shine in the last 2,000 years.

These six or seven stars and some two hundred more nearby are closely related. They travel together and are all at about the same distance—400 light years—from us. That's what earns them the classification of cluster.

Just because we group stars into one constellation or asterism does not make them related. The stars in the Big Dipper, the asterism many people use to find Polaris, prove that.

From the measured motion of each of its seven brightest stars, we can tell that the Dipper looked very different 50,000 years ago and will look quite different again in another 50,000 years. Definitely not a cluster.

The Pleiades are called an open cluster to distinguish their loosely associated stars from globular clusters. To the naked eye the nearest one may look like a single star. In a small telescope the star may show slightly fuzzy edges.

With one of the largest telescopes you'd see an enormous number of stars—tens of thousands or more—in a compact spherical space. Compact, that is, when you consider the number of stars in it. The Hercules cluster, M13—difficult to find and barely visible to the naked eye, 22,000 light years from us—is 300 light years across.

About 130 globular clusters have been identified in our galaxy. Such clusters also have been found in other galaxies, for example, the Andromeda nebula.

Variable Stars

You can't see double stars without a telescope. When you see two stars close together, they just happen to be near each other when seen from Earth. The "rider" atop the middle star in the handle of the Big Dipper is an example.

True binary stars are linked by gravity, like the moon and Earth, or Earth and sun. They seem to be very common. It takes only a 3-inch (75 mm) telescope to show the companions of Polaris, Rigel, Aldebaran, Antares, or Sirius.

Astronomers estimate that one out of every three stars has a companion. Stars with several companions are much rarer. Castor has two companions, and one of them is itself a binary.

Binary star systems explain some of the variable stars, stars that change in magnitude. Algol (AL-gol), shown in Figure 9.1 as an open circle near 3^h, is the classic example.

Normally of magnitude 2.3, it drops within five hours to magnitude 3.5 when the dimmer star eclipses the brighter one. Five hours later, when we again see both stars, Algol recovers its normal brightness. Its name, which translates into "demon," suggests that the Arabs had noticed the star's strange behavior centuries ago.

Betelgeuse is variable for a different reason. It changes irregularly, between magnitudes 0.1 and 1.2, depending on its size, which is controlled by its atomic furnace.

Other stars change brightness with clocklike regularity in periods from a few hours to a few months or even years.

Here is the explanation: The star contracts and thereby heats up until a new atomic process becomes possible. That releases new energy, which makes the star expand. When expansion makes the distance between atoms too large the reaction stops. The outer layers of the star cool. Gravity pulls the star together again, and the cycle starts over.

These slowly growing and shrinking—pulsating—stars are not to be confused with pulsars that send radio pulses at regular intervals of seconds or fractions of seconds.

"New" Stars

Occasionally a star seems to appear out of nowhere. It's a nova, Latin for "new." We now know that name to be wrong. There has been a dimmer star where we now see the brighter one. When a star increases ten magnitudes but is still visible only in institutional-size telescopes, that interests only astronomers.

Very occasionally a star in or near our galaxy explodes and becomes so bright it attracts worldwide attention even from nonastronomers. In November 1572 such a supernova flared up in the constellation Cassiopeia, virtually in view of the astronomer Tycho Brahe. Brighter at first than Jupiter at its brightest, it soon became brighter than Venus, and finally was easily seen in daylight. It gradually dimmed, and Tycho lost track of it in March 1574.

Before that, a star flared up in Taurus in 1054. The Crab nebula, M1, named for its shape, can still be seen in telescopes. It is the expanding shell of that explosion and still sends x-rays to radio telescopes on Earth.

Not long after Tycho's discovery, Kepler—his assistant and successor—discovered a supernova in March 1604. It was observed with the naked eye until March 1606, before the telescope was invented.

No supernovae for 383 years.

Then on February 24, 1987, Ian Shelton, a 29-year-old Canadian, repaired an old telescope at the Las Campanas mountain-top observatory in Chile. To make sure it was in working order he took a photograph of the Large Magellanic Cloud, the galaxy nearest ours—invisible in our area.

At a distance of 170,000 light-years from us it shows a lot of nebulosities but no bright stars. When—after a three-hour exposure—he developed the plate, a large round blotch showed up. His first thought: a flaw in the emulsion.

But flaws on the plates used by astronomers are rare. And they are not that big. "It can't be an exploding star," he kept telling himself. By coincidence he had taken a picture of the same area the night before. Not a sign of a blotch there.

Shelton stepped outside. Next to a whispy patch called the Tarantula nebula, in Dorado (the Swordfish or Goldfish) he saw a star where yesterday had been none.

You and I would not have been impressed with that fifth-magnitude star. But Shelton, an astronomy buff, had worked the last four lonely years out of six on this mountain in the Andes. Nursing the telescopes of the University of Toronto was his main job.

He knew exactly what that dot of light meant. Overnight, one hundred seventy thousand years ago a star had exploded near 30 Doradus. A supernova. It might be the first nearby supernova since Kepler's star. If nobody had seen it and tele-graphed the news before he did, it might become *his* supernova.

It is now known as Supernova Shelton.

Comets

Chances of your seeing a spectacular supernova in your lifetime are, judging by the past, slim. Chances of seeing a bright comet are much better.

Every year, a half-dozen or more comets can be seen in telescopes. Every ten years one is readily visible to the naked eye if you know where to look. Within a human lifespan, perhaps one may become impressive.

Astronomers still don't have all the facts about comets, but they know their general structure. The nucleus, a few miles or kilometers across, is made up of solid particles: loose rocks, ranging in size from boulders to dust, and frozen gases.

Most comets we see are permanent members of the sun's family. Other family members, the major and most minor planets, move in almost circular ellipses, and hug the plane of the ecliptic. (Pluto, the outermost known major planet, thumbs his nose at both these rules.)

Comets move in elliptical orbits, but the ellipses resemble cigars more than circles. These comets return again and again to the vicinity of the sun. Others approach the sun, turn, and seem to leave the solar system forever. The orbits of both the repeaters and the one-time visitors make odd angles with the plane in which the planets move.

Only when a comet comes as close to the sun as, say, Saturn does it become visible. Its ices then form clouds of gases around the nucleus. This gas cloud, the size of Earth or up to ten times as large, reflects sunlight. Radiation from the sun causes these gases to glow and give off more light.

The solar wind—protons and electrons thrown out by the sun—presses on these gases. That makes them stream out behind the head of the comet in a tail, which can be a hundred million miles long.

All this time the comet, held together weakly by its small gravity, loses material that scatters along its orbit. The comet will be brightest when it's nearest the sun, but then it's lost to us in the glare. A comet will be seen at its brightest just before it gets lost in the glare of the sun, or when it emerges from it. That means you'll see it shortly before dawn or after dusk. At these times, it also means the comet will be near your horizon, just as it does for Mercury. Brighter comets can be seen higher in the sky, morning or evening, like Venus.

Most comets are discovered, or rediscovered on their return— often by amateur astronomers—long before they become visible to the naked eye. That gives publications such as *Sky and Telescope* time to publish predictions. Such predictions will be in units you now know well: right ascension, declination, and magnitude for some dates in the future.

You can pencil in these positions on one of the maps in this book, say Figure 9.2. The first time you do that, don't forget that RA here increases toward the left. Declination everywhere increases away from the equator.

For naked-eye observation, later predictions may show a star map of the section of sky where you should search for the comet from week to week.

The only problem you'll have with the predictions is the magnitude. Comet Kohoutek, forecast to become the comet of the century—visible in daylight in 1974—turned out to be hard to find even with binoculars.

If you can't find the comet, cheat. Borrow a pair of binoculars—7 × 50s will do nicely. Cover the area by sweeping the sky just above the horizon with the glasses. Raise them a little and sweep back. Raise the glasses a bit more and repeat.

There's little danger of confusing the comet you are looking for with a planet or star. Unlike these bodies the comet will have some shape to it. When its tail is short or dim, the comet will look like a fuzzy grain of rice.

Of the returning comets, many show up every three to seven years. On each round trip they lose some of their gases until they become impossible to see for casual observers.

Most of the longer-period returning comets recede far from the solar system between appearances. Their last close approach to the sun may have been before Gauss (1777 to 1855) found a way to calculate their orbits.

One exception is the best known of all comets: Halley's (HAL-lees, not HAYL-lees). It only retreats to about the distance of Pluto. It's named for the English astronomer who first suggested that the comet seen in his lifetime in 1682 was the comet that had been recorded in 1531 and 1607. He predicted its return in 1759. And it did return.

It has been traced back 3,000 years with appearances about every 76 years, through Chinese and Babylonian records among others. In 1910 it was not as bright as had been expected, my father told me. In 1987 it was barely visible to the naked eye for a few weeks. Is it running out of gas?

Even though you can't see comets very often, you can see some of the stuff they were made of several times a year.

Meteors

Every clear night you see shooting stars. Many of them are bits of matter that once were part of a comet. When they get into the atmosphere—say 65 miles (100 km) above the ground—they heat up by friction and begin to glow. Most of them burn themselves out before they reach the 40-mile (65 km) level.

A rare larger meteor may explode as a fireball, visible over hundreds of miles. Occasionally one is large enough to survive the heating on the way down and reaches the ground as a meteorite.

You'll see more shooting stars after midnight than before. That's when your part of Earth is facing forward in our planet's orbit around the sun. So your side collects more hits, just as the windshield of a moving car gets more raindrops than its rear window.

On some nights you see not only the usual number of such "sporadic" meteors, but many more. They are almost certainly leftovers from a comet, gravel spilled along its orbit.

Earth recrosses such fields of comet debris on almost the same day each year.

That lets us predict when such meteor showers can be expected. Between October 16 and 24, for instance, Earth crosses the orbit of Halley's comet. Every year around October 20 you can see some fireworks in the sky. During the four days—the spread in the accompanying table—before and after the main event, you'll see a lesser display.

That shower is called the Orionids. Why? If you plotted on a star map the path of the meteors you see during any of these nights, the majority would seem to radiate from a point near RA 6.2^h, DEC 15°, which is in the constellation Orion.

How many meteors you'll actually see is unpredictable. One year a shower that usually produces twenty shooting stars per hour may have several hundred. The next year, also in the dark of the moon and after midnight, you may not count more than ten.

On the next page is a table of fairly reliable annual meteor showers.

Name of Shower	Date of Maximum	Spread Days
Quadrantids	Jan 3	1
Lyrids	Apr 20	1
Aquarids	May 4	9
Aquarids	Jul 29	10
Perseids	Aug 11	2
Orionids	Oct 20	4
Leonids	Nov 16	2
Geminids	Dec 13	3

On these dates and others, many amateur astronomers are out observing, counting, and recording meteors. The main requirements: patience, insect repellent, and warm underwear.

Summary

- From the phase of the moon you can predict periods of darkness.

- During such dark periods you can see dimmer objects in the sky: the Milky Way, nebulae, and comets.

- On some nights, in the dark of the moon and especially in the last hours of the night, you'll see far more than the usual number of shooting stars—a meteor shower.

II

Worldwide Sky Watching and Direction Finding

12

The North Temperate Zone

In Part I you looked at stars and planets as seen from mid-northern latitudes, say the touching forty-eight states or most of Europe.

But people take cruises and they vacation in other parts of the world. They fool the seasons at home by skiing in South America in August or sailing in Australian waters in December. They work on Alaska's North Slope, and man weather stations in the Antarctic.

For these people Part II expands the coverage of this book to the entire world, from North Pole to South Pole. It's also for hammock readers who wonder what's different in the sky over far-off places.

In Part I we barely touched how you can use your knowledge of the heavenly bodies. In Part II you'll read about finding North or South, East or West.

The methods that let you do these things vary from one part of the world to the next.

Around 140 AD, the astronomer Ptolemy of Alexandria divided the Earth into three climate zones: torrid, temperate, and frigid.

The basic idea goes back another three hundred years to Eratosthenes. It holds that the climate of an area depends on the angle (*klima* in Greek) of sunshine.

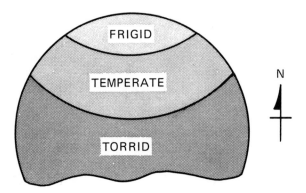

12.1   The Climate Zones of Ptolemy.

In the torrid zone—the Tropics—on either side of the equator, the sun is close to being overhead all year long. For discussing sky watching at night one has to divide that zone into the parts north and south of the equator.

There are two temperate zones—one north, one south of the tropical zone. In either zone the sun rises and sets every day of the year. At noon it is always at a moderate angle above the horizon, never directly overhead.

One of the frigid zones, the Arctic, lies north of the North Temperate Zone. The other, the Antarctic, lies south of the South Temperate Zone. In both frigid zones the sun on some days does not rise at all, on others it stays up all night.

We know now that the angle of sunshine is only one factor in the climate of an area. But the zones of Ptolemy and their boundaries are still important for sky watching and for finding direction.

Zones are the most logical way for answering questions such as: Why is the moon in some areas invisible two weeks out of four? Why does a first-quarter moon look like a last-quarter moon in another area? Why are Jupiter and Saturn absent from your sky for years on end? Why do familiar constellations stand on their heads?

The North Temperate Zone (NTZ) stretches from the Tropic of Cancer to the Arctic Circle, or from about 23½°N to 66½°N. From border to border that's about 3,000 statute miles (4,800 km), a little less than half the distance from equator to pole.

In either temperate zone, the length of day varies more than in the torrid zone, but less than in the frigid zones.

Wherever you are in the NTZ, you'll be aware of the length of day—the time from sunrise to sunset—changing with the season. Already at the southern boundary of the NTZ it varies from 10¾ hours in midwinter to 13½ hours in midsummer. A difference of about three hours.

In the center of the NTZ, near latitude 45°N, the length of day varies from 8¾ hours in midwinter to 15½ hours in midsummer. The difference has grown to about seven hours.

At the northern boundary of the NTZ, near latitude 66½°N, the midwinter rise and set times are uninteresting. Even at noon the sun gets only one half-finger above the horizon. The rather theoretical length of day then is 2¼ hours. In midsummer at that boundary the sun never sets: midnight sun. So the difference between the shortest and longest day here is almost 22 hours.

At noon on the longest day at the north boundary, the sun is 47° above the horizon. So the sun's noon altitude here varies during the year from near 0° to 47°.

More typical of the sun's angle above the horizon at noon in the NTZ is the middle of the zone. In latitude 45°N the angle varies from 21½° in midwinter to 68½° in midsummer.

Geography

You already have a pretty good idea what areas are in the NTZ. Here are a few fine points:

- All of the touching 48 states of the U.S.A. are in this zone. Key West, Florida, misses the tropical zone by one degree; Brownsville, Texas, misses by 2½°.

- Only a small part of Alaska—small by Alaskan standards— including the Brooks Range and the area north of it are north of the NTZ, in the Arctic.

- The Hawaiian Islands straddle the Tropic of Cancer. All the larger, better known islands are in the Tropics. Only the

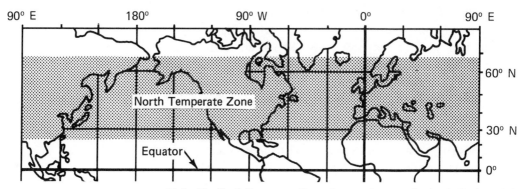

12.2   The North Temperate Zone, the area between the Arctic Circle and the Tropic of Cancer.

smaller northern islands—from Necker to Midway—are in the NTZ.

• Canada's ten provinces from Newfoundland to British Columbia are entirely in the NTZ. So are large parts of the Yukon and the Northwest Territories.

• Mexico's Baja California is in the NTZ except for its very tip. On the mainland a line roughly from Mazatlán to Ciudad Victoria forms the southern boundary of the zone. No part of Central America is in the NTZ.

• The Bahamas that most people visit are in the NTZ. The Tropic of Cancer cuts through Great Exuma (a little south of Georgetown) and the northern tip of Long Island.

• All the Caribbean islands, from Cuba to Trinidad, are outside the zone.

• Greenland's southern tip and all of Iceland are in the NTZ.

• Europe is solidly in the NTZ. The exceptions are the northernmost 300 miles (500 km) of the Scandinavian peninsula that includes parts of Norway, Sweden, and Finland, and the adjoining Kola peninsula of the USSR. The Arctic Circle continues to the Urals—which divide Europe from Asia—and cuts off the northernmost Russian land area and offshore islands.

- North Africa, not just the shore of the Mediterranean but deep into the continent—say Egypt to the Aswan dam—lies in the NTZ.

Fixed Stars in the NTZ

The celestial North Pole in the NTZ is as many degrees above your northern horizon as you are north of the equator. For example, at the southern limit of the North Temperate Zone, latitude 23.5°N, the celestial North Pole is 23.5° above the point due north of you on the horizon.

The other pole, the celestial South Pole, is as many degrees below your southern horizon, hence invisible.

The celestial equator touches the East and West points on your horizon. It rises 90° minus your latitude degrees above your South Point. For example, in latitude 23.5°N, the high point is 66.5° above your South Point.

Stars in or attached to the North Cap never set if their declination equals at least 90° minus your latitude.

For example, in latitude 23.5°N stars with a declination of 66.5°N or greater always stay above the northerly horizon. Here none of the stars that form the Big Dipper qualify. Dubhe, the Pointer nearer the pole, is in declination 62°N.

Stars near the bottom of our strip map, say Figure 9.2, or in the southern gap will never rise if their south declination is greater than 90° minus your latitude.

12.3   The dome of the sky over NTZ, shown by the north polar cap and the strip map in place. Left: at the Arctic Circle. Right: at the Tropic of Cancer.

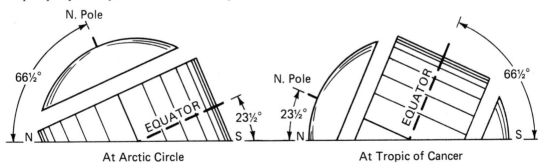

At Arctic Circle   At Tropic of Cancer

For example, in latitude 23.5°N only stars in declination 66.5°S or less ever appear above the horizon. Acrux, here at the foot of the Southern Cross, is in declination 63°S. That makes the whole Cross rise a little above the horizon.

At the northern limit of the North Temperate Zone, in latitude 66.5°N, the celestial North Pole is 66.5°—more than three spans—above the point due North of you on the horizon.

The celestial South Pole is as many degrees below your southern horizon, hence invisible.

The celestial equator here—as everywhere in the world—touches the East and West points on your horizon. It rises to 90° minus your latitude degrees above your South Point. In latitude 66.5° that is 23.5°.

Stars in the northern sky never set if their declination equals at least 90° minus your latitude. In latitude 23.5° that's at least 66.5°. That includes all the stars in the northern cap; the nearby stars Capella, Vega, and Deneb; and—from the strip maps proper, say Figure 9.2—Pollux.

Stars near the bottom of the strip map or in the southern gap will never rise if their south declination is greater than 90° minus your latitude. At the edge of the Arctic, only stars with south declinations of 23.5° or less ever appear above the horizon.

We lose Antares and Fomalhaut there. You'd have to travel about 200 miles south of the Arctic Circle to see Antares on the horizon, 450 miles to see Fomalhaut.

The planets and the moon occasionally get a few degrees south of the southernmost point of the sun's path in declination 23½°. If you are close to the northern border of the NTZ you will lose sight of them at these times. For example, in October 1999 Mars gets to declination 25°S. Appendix B may give you advance notice of that loss.

Finding Directions in the NTZ

A Star to Steer By

When the stars are out you can steer your boat without constantly watching the compass. That's the celestial equivalent of steering for a lighted buoy after checking with the compass that it's the right one.

The technique is simple. You put the boat on the proper course by compass. Then you find a star directly ahead. A space between two stars will do just as well. Your neck will be happiest if you pick a star or spot a span or so above the horizon. You don't have to know the name of the star or its constellation; a planet works just as well.

For the next few minutes steer toward the star. Then glance at the compass. If you are still on course continue for a few more minutes. Then check the compass again.

Unlike buoys, stars don't stay put. Before long you'll find the star a bit to one side of the correct course. Bring her on course by compass, then continue aiming at the empty spot a little way from the guiding star.

If you have chosen a space between two stars, gradually aim farther from one of the stars and closer to the other. When you have to steer too far from your original mark find a new one, perhaps one a little higher or lower. You can repeat that game until dawn.

I avoided saying the star had moved toward the west. In the area between the northern Tropics and the Arctic, most stars seem to move from East to West by way of South. All the stars on our strip map and in the southern gap (both in Figure 3.2) do that. But stars in the polar cap may at times appear to drift eastward.

How fast do the stars drift? Or, more practically, how often should you check? That depends in a complex way on your latitude and longitude and the star's right ascension and declination and the star time at the moment.

It's not something you can calculate in your head. Here are some figures that let you estimate the drift:

• In the zone from the Arctic Circle to the Tropic of Cancer, a star one span (20°) above the horizon will on most courses seem to drift about one finger (2°) in 12 minutes.

• On southerly courses it'll drift about one finger in 8 to 10 minutes.

You'll probably find watching a star more pleasant than staring at the compass all the time. And you get away from the hypnotic effect of the compass.

You can borrow that helmsman's trick for walking at night, in the desert. Here it is the astronomical equivalent of what an experienced hiker would advise you to do in daytime. "Find a mark dead ahead on the wanted compass course, an odd-shaped rock, perhaps, or a withered bush.

"Then walk toward that mark. Ignore the compass; use your eyes to watch your step. When you reach the first mark, use the compass to find a new rock, and so on to your destination."

To adapt the helmsman's trick to land navigation, set the compass for the desired course from the map. Then find a star on the same bearing at that time.

Again the star—or the space between two stars—or a planet should be comfortably close to the horizon. After walking a few minutes, check the star with the compass. Continue toward it, if it's still close enough to your course. If it isn't, aim a little to one side of it. Most likely you'll be walking a more easterly course. Only on a northerly course may you ever have to correct by veering a little westward.

A whole book can be written about modern map and compass work, including what compass to get. Also how best to deal with magnetic declination because magnetic compasses hardly anywhere point north. (In Maine, for example, compasses may point 20° west of north, in Washington State 20° east of north.) I have written such a book for Sierra Club Books. Its title is *Land Navigation Handbook*.

For general work I suggest a plate compass, also called a protractor compass, with mirror sight and declination adjustment. Two popular makes are Silva® and Suunto®. They come with instructions to get you started taking courses off the map, and taking bearings, say of a distant peak. Here you use a star as target in place of a peak.

Say you want to cross a hard-frozen lake at night. Using the map, find a landing site on the far shore. With your compass get the course to walk, ski, or snowshoe. Find a guide star. From time to time check the star's drift and correct your course. When the star gets too far off, find a new star.

Polaris

Everywhere in the NTZ, the North Star, Polaris is above the horizon at all times. Its name translates into Pole Star, a logical name for a star less than half a finger's width from the celestial North Pole, the point above the North Pole on Earth.

That's the point where the upper end of the shaft comes out on a schoolroom globe. On the real Earth that shaft is not fixed in space but traces a circle in the sky. The radius of that circle is more than two spans. It takes some twenty-six thousand years to complete one sweep.

So the axis around which Earth rotates did not always point toward our Pole Star. Alpha in Draco, the Dragon, about three hands from Polaris, marked the celestial pole for the builders of the earliest pyramids in Egypt.

In the North Temperate Zone, in Egypt or in the middle of the Atlantic, a marker at the celestial North Pole shows exactly where North is. Polaris only now marks that spot accurately enough for backpacking in the wilderness or laying out a flower bed. Land surveyors can see Polaris in daylight through their telescopes; they then correct for the small distance Polaris misses being exactly at the pole at the time of observation.

For naked-eye work, consider the bearing of Polaris as true North.

If for some reason you want to find North with greater precision, do so when Polaris is directly above or below the celestial pole. That is when your local star time is $2\frac{1}{2}^{\text{h}}$ or $14\frac{1}{2}^{\text{h}}$. Most of the year one of these times will fall into the time of darkness—over most of the area.

For the next hundred years the small error gets even smaller. The name North Star is now very apt. Once you know in what direction North lies, you can work out all other directions.

To translate a clock face into a compass card imagine, or draw, this:

> North (0° or 360°) at 12 o'clock
> East (90°) at 3 o'clock
> South (180°), opposite North, at 6 o'clock
> West (270°), opposite East, at 9 o'clock

Each hours adds 30°, so
1 o'clock is 30°,
2 o'clock is 60°,
and so on until
11 o'clock is 330°.

Polaris and the Strip Map

To find Polaris from the Big Dipper, start at the back edge of the bowl. Trace that line away from the rim for one span four fingers. You'll arrive near second-magnitude Polaris, with no star that bright within eight fingers. No planet can ever confuse you in that part of the sky.

Polaris can do more than just show you where North is:

• You can get your latitude from it. The rule is simple: The distance between Polaris and the horizon below it equals your latitude. (With a mountain obstructing your horizon, estimate where the horizon would be if you could move the mountain out of the way.)

If, for example, you get two spans, which equals 40°, you are near latitude 40°N.

• You can also use Polaris to find the high point of the equator south of you. With Polaris directly behind you, true South is directly in front of you. Just measure 90° minus your latitude (which in the example gives 50°) up from your south horizon.

• Next give the point just found a Roman salute as in Figure 2.3. Then swing your arm left to the East Point on the horizon; bring it back to the high point and continue the swing to the west point on the horizon.

Your hand has just traced the celestial equator, as in Figure 2.5.

That alone will often let you match the equator of the strip map to the one in the sky. Just mentally pull the strip map sideways until one feature in the sky—one star, constellation, or super-constellation—matches the sky.

But suppose you can't match the Hexagon or any of the triangles. Perhaps patchy clouds make things tough. There are several ways for getting star time.

• You could look up approximate local star time from date and civil time in Appendix A.

• You could turn the polar star map (Figure 6.4) until you get a match with the part of the northern sky you can see. After matching map to sky, the number on top is the current star time. That time needs no corrections for longitude, or daylight saving time.

• Unless you are a mathophobe, you could also roughly calculate the present star time on a match folder or in your head.

• You can also get rough star time from Polaris and its nearest equally bright neighbor, a hand and three fingers away. That star is Kochab (beta in Ursa Minor, #40). There are no nearby stars to confuse you, and planets never get to that area.

Consider the line from Polaris to Kochab as the hour hand of a clock centered on the celestial North Pole. When Kochab seems directly above Polaris, star time is within a few minutes of 15^h. Directly below Polaris it shows 3^h with the same accuracy.

Star time found from that clock needs no correction for daylight saving time. Nor do you have to allow for the difference in longitude between your location and the standard time meridian of your time zone.

This clock is visible, all night, and every night, everywhere in the NTZ. It sounds great to have a star clock, visible in most of the northern hemisphere at any hour of darkness.

If you only need star time when Kochab is in the 12 o'clock or 6 o'clock position, that's fine. At all other times that clock is not reliable.

Since 15^h is on top you expect 21^h, six hours later, to be at the same height above the horizon as Polaris. In the middle of the NTZ, in latitude 45°N, your estimate would be about three quarters of an hour off. Near the Arctic Circle the error is one hour and a quarter.

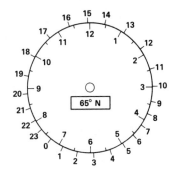

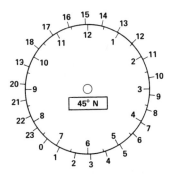

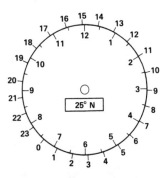

12.4 Polaris/Kochab star-time dials for different latitudes in the NTZ. Read the position of Kochab as the end of the hour hand of a clock that has Polaris at the center of its dial. With the time shown inside the circle, go to the dial nearest your latitude and read star time on the numerals outside the circle.

Example: Kochab, at latitude 45°N, is at 4 o'clock. Star time is 7ʰ.

Here are three diagrams showing how to get local star time from Kochab and Polaris more accurately.

You may wonder why Kochab's hour hand doesn't move exactly six star hours in three civil time hours.

Common sense is certainly on your side. The Kochab hand sweeps out—close enough—24 hours, while we read the clock hand sweep as having moved 12 hours.

Common sense and some writers of books overlook a detail. Other clock dials are flat and near eye level. But we read the Polaris–Kochab dial high up on the underside of a spherical dome.

In place of Kochab you can use the Pointers (page 54 and Figure 6.2) and Polaris as the hour hand. Near latitudes 25° and 45° simply subtract 4 hours from the star time shown on the outer dial. Example: The Pointers, at latitude 45°, are at 4 o'clock. Star time is 3ʰ. Farther north the simple Pointers and Polaris clock becomes inaccurate.

Directions from Special Stars

At the beginning of this book we fitted the star map to the sky. We made the horizontal zero line, the celestial equator, touch the horizon exactly East and West of the observer. The observer's latitude had nothing to do with it.

The stars have been informed of our procedure. You'll find that the stars cooperate and appear just where our model predicts.

A star on the celestial equator (with zero declination) will go through the point due East of you on the horizon when it rises. It will go through the point due West of you on the horizon when it sets. Your latitude has nothing to do with it.

There's another way to arrive at the last three statements. Everywhere in this zone in winter, the sun rises south of east and sets south of west. In summer the sun rises north of east and sets north of west. At the change of seasons—about March 20 and September 22—the sun crosses the celestial equator. On these days, everywhere in this zone, the sun rises due East and sets due West. Again, latitude has nothing to do with it.

If we can find a star that's where the sun is on these days— on the celestial equator—we can find true East whenever that

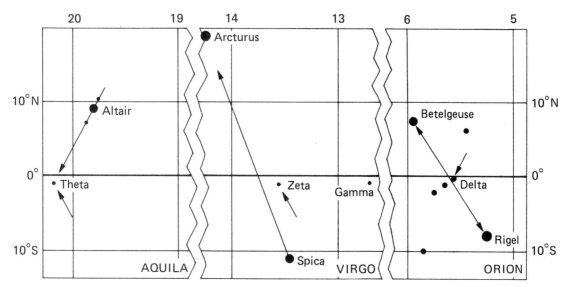

12.5   Equatorial stars rise everywhere due East, set due West.

star rises, true West when it sets. Since stars don't noticeably change declination, we can do that any night of the year, anywhere in this zone.

No first-magnitude star is really on the equator. But stars in three constellations can serve as a compass at rise or set. They are Orion, Virgo, and Aquila, all readily found from their first-magnitude stars.

In Orion the star nearest the equator is delta, Mintaka, the westernmost of the stars that form the belt. Its neighbor, the middle star—epsilon, Alnilam (#15)—is only about half a finger's width from the equator, close enough for finding East and West.

In Virgo, of two possible stars, zeta very close to the equator is the easier one to find. It is roughly one hand north of Spica and two fingers west of the line that connects the first-magnitude stars Spica and Arcturus. If it's dark enough for you to see Virgo's shape (Figure 5.8), zeta is the end of her belt away from Spica. Gamma, about as close to the equator as zeta, is the maiden's neck.

In Aquila the star nearest the equator is theta, a little more than a hand from first-magnitude Altair. Altair and its two

Star Time(^{h})	Constellation	Phenomenon
2¼	Aquila	theta sets
6¾	Virgo	gamma rises
7½	Virgo	zeta rises
11½	Orion	delta and epsilon set
14¼	Aquila	theta rises
18¾	Virgo	gamma sets
19½	Virgo	zeta sets
23½	Orion	delta and epsilon rise

distinctive dimmer outriders point to it in a roughly southerly direction.

It is easier to identify a setting star than a rising one because you'll have had time to study the entire constellation. The table above gives approximate star time of rise and set.

It's a rare night when you don't have at least two constellations rising or setting. But sometimes moonlight interferes; at least Orion is moonproof.

At sea, on a large lake, or on flat land—whatever your elevation above sea level—the horizon is no problem in naked-eye observation. Where hills or distant mountains hide the East or West point, the bearing at rise will be a bit south of East, at set a bit south of West. How much south will depend on your latitude.

Near the tropics the bearing of these stars changes one finger's width in the 25 minutes after sunrise, and before sunset. Near the Arctic Circle it changes that much in 12 minutes.

Direction from the Sun

Why mess with second- and third-magnitude stars for finding direction when we can use a star of magnitude minus 27, one that everybody recognizes? The sun.

At Sunrise and Sunset

Unlike stars, the sun changes declination during the year through a range of no less than 47°. Only at the equinoxes—when the sun's declination is zero—does it rise and set on the same bearings everywhere. On all other dates the bearings at rise and set depend on the latitude of the observer.

For two hundred years navigators have checked their ships' compasses twice a day, at sunrise and sunset. The two tables following let you get these bearings from the Tropics to latitude 54°N.

You can do the same. That'll be handy when you don't know the local compass error (declination or variation), or when you are in doubt whether to add or subtract it.

You may arrive at a possible campsite just before sunset. Then you could work out tomorrow's route tonight. If you arrived too late, you could work it out early next morning.

Don't fret if sunset is hidden by an obstruction on the horizon, or if you missed set or rise by a few minutes. At these times the bearing of the sun changes rather slowly. For example, it changes a single degree in 9 minutes in latitude 25°; in 6 minutes in latitude 40°; and in 5 minutes in latitude 54°, the limit of the tables.

All the data you need for a long hiking trip will fit on a wallet card. All you'll probably need is two columns for latitude and a few lines for the dates of your trip. So you don't have to carry this book with you just for these tables.

The Sun at Noon

Everywhere in this zone the sun bears due South when it is highest in the sky, that is at noon. That sounds like a good time to get true South from the sun.

The weather may have been too awful to start in the morning. Or you'd like to check that you are still on the right course. A few people have been known to lose their compasses, and even themselves, in the forenoon. All good reasons for checking your direction from the sun at noon.

Only by chance will noon at any place ever be at 12:00.

The Earth orbits the sun not in a circle but an ellipse. The axis of Earth is inclined to its orbit. These two facts combine to shift noon up to about a quarter hour before or after twelve o'clock local time.

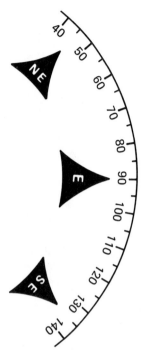

How to use the sunrise and sunset tables: 1. Find the line nearest today's date. 2. Scan across the line to the column with the latitude nearest yours. 3. Where the line and column cross, read the true bearing of the sun. Example: *What is the bearing of sunrise May 8, in latitude 40°N?* Answer: *68°.* Second example: *What is the bearing of sunset, November 15, in latitude 36°N?* Answer: *247°. The compass cards in the margin let you relate the figures in the tables to more familiar compass bearings.*

Bearing of Sunrise (in degrees)

Date		25°	34°	Latitude, North or South 40°	44°	47°	50°	52°	54°
Jan	1	116	118	121	123	125	127	129	132
	16	113	116	118	120	122	124	126	128
	25	111	113	115	117	119	120	122	124
Feb	2	109	111	112	114	115	117	118	120
	8	107	108	110	111	112	114	115	116
	15	104	106	107	108	109	110	111	113
	20	102	103	104	105	106	107	108	109
	26	100	101	102	103	103	104	105	105
Mar	3	98	98	99	100	100	101	101	102
	8	96	96	97	97	97	98	98	99
	13	93	94	94	94	94	95	95	95
	18	91	91	91	91	91	92	92	92
	23	89	89	89	89	89	88	88	88
	29	87	86	86	86	86	85	85	85
Apr	3	84	84	83	83	83	82	82	81
	8	82	82	81	80	80	79	79	78
	13	80	79	78	77	77	76	75	75
	19	78	77	76	75	74	73	72	71
	25	76	74	73	72	71	70	69	67
May	1	73	72	70	69	68	66	65	64
	8	71	69	68	66	65	63	62	60
	16	69	67	65	63	61	60	58	56
	26	67	64	62	60	58	56	54	52
Jun	10	64	62	59	57	55	53	51	48
	30	64	62	59	57	55	53	51	48

Only on four days each year is the sun due South at 12:00 on a sundial, designed for its location and correctly set up. These days fall in the middle of April and June, at the beginning of September, and in late December.

To get correct time from a sundial you need a calendar that shows how many minutes the sun is fast or slow on local mean noon at any given day. The following table is such a calendar. It lets you set up a sundial correctly without waiting for one of these four days.

More usefully, the table also helps you get true South from

Bearing of Sunrise (in degrees)

Date		25°	34°	Latitude, North or South 40°	44°	47°	50°	52°	54°
Jul	1	64	62	59	57	55	53	51	48
	19	67	64	62	60	58	56	54	52
	28	69	67	65	63	61	60	58	56
Aug	5	71	69	68	66	65	63	62	60
	12	73	72	70	69	68	66	65	64
	19	76	74	73	72	71	70	69	67
	25	78	77	76	75	74	73	72	71
	30	80	79	78	77	77	76	75	75
Sep	5	82	82	81	80	80	79	79	78
	10	84	84	83	83	83	82	82	81
	16	87	86	86	86	86	85	85	85
	21	89	89	89	89	89	88	88	88
	26	91	91	91	91	91	92	92	92
Oct	1	93	94	94	94	94	95	95	95
	6	96	96	97	97	97	98	98	99
	11	98	98	99	100	100	101	101	102
	17	100	101	102	103	103	104	105	105
	22	102	103	104	105	106	107	108	109
	28	104	106	107	108	109	110	111	113
Nov	3	107	108	110	111	112	114	115	116
	10	109	111	112	114	115	117	118	120
	18	111	113	115	117	119	120	122	124
	27	113	116	118	120	122	124	126	128
Dec	11	116	118	121	123	125	127	129	132
	31	116	118	121	123	125	127	129	132

the sun at noon on any day and at any place between the Tropics and the Arctic. But other corrections are needed.

It's a safe bet that your watch or clock is not set to local time. Most likely it is set to some zone time, say Eastern Standard Time (five hours slow on Greenwich Mean Time, now officially called Universal Time). Or perhaps your watch is set to Central European Time (one hour fast on UT).

In summer you may be on Eastern Daylight Time, one hour ahead of standard time. Noon Eastern Standard Time is near one o'clock by your watch.

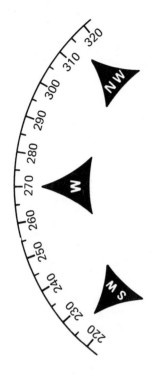

Bearing of Sunset (in degrees)

Date		25°	34°	40°	44°	47°	50°	52°	54°
					Latitude, North or South				
Jan	1	244	242	239	237	235	233	231	228
	16	247	244	242	240	238	236	234	232
	25	249	247	245	243	241	240	238	236
Feb	1	251	249	248	246	245	243	242	240
	8	253	252	250	249	248	246	245	244
	14	256	254	253	252	251	250	249	247
	20	258	257	256	255	254	253	252	251
	25	260	259	258	257	257	256	255	255
Mar	3	262	262	261	260	260	259	259	258
	8	264	264	263	263	263	262	262	261
	13	267	266	266	266	266	265	265	265
	18	269	269	269	269	269	268	268	268
	23	271	271	271	271	271	272	272	272
	29	273	274	274	274	274	275	275	275
Apr	3	276	276	277	277	277	278	278	279
	8	278	278	279	280	280	281	281	282
	13	280	281	282	283	283	284	285	285
	19	282	283	284	285	286	287	288	289
	25	284	286	287	288	289	290	291	293
May	1	287	288	290	291	292	294	295	296
	8	289	291	292	294	295	297	298	300
	16	291	293	295	297	299	300	302	304
	26	293	296	298	300	302	304	306	308
Jun	10	296	298	301	303	305	307	309	312
	30	296	298	301	303	305	307	309	312

Eastern Standard Time is the local mean time in longitude 75°. Most time zones stretch 7½° east and west from their reference longitudes. Halfway between the Tropics and the Arctic, that's a stretch of 750 miles (1200 km) in all. Noon is one whole hour later at the western limit of the time zone than the eastern.

To get the time of noon accurate enough to tell when the sun is due South of you, you must know your location within a few miles or kilometers. Then you must have a detailed map of your area from which to read your longitude.

Bearing of Sunset (in degrees)

Date		25°	34°	Latitude, North or South				52°	54°
				40°	44°	47°	50°		
Jul	1	296	298	301	303	305	307	309	312
	19	293	296	298	300	302	304	306	308
	28	291	293	295	297	299	300	302	304
Aug	5	289	291	292	294	295	297	298	300
	12	287	288	290	291	292	294	295	296
	18	284	286	287	288	289	290	291	293
	24	282	283	284	285	286	287	288	289
	30	280	281	282	283	283	284	285	285
Sep	4	278	278	279	280	280	281	281	282
	10	276	276	277	277	277	278	278	279
	15	273	274	274	274	274	275	275	275
	20	271	271	271	271	271	272	272	272
	26	269	269	269	269	269	268	268	268
Oct	1	267	266	266	266	266	265	265	265
	6	264	264	263	263	263	262	262	261
	11	262	262	261	260	260	259	259	258
	16	260	259	258	257	257	256	255	255
	22	258	257	256	255	254	253	252	251
	28	256	254	253	252	251	250	249	247
Nov	3	253	252	250	249	248	246	245	244
	9	251	249	248	246	245	243	242	240
	17	249	247	245	243	241	240	238	236
	26	247	244	242	240	238	236	234	232
Dec	11	244	242	239	237	235	233	231	228
	31	244	242	239	237	235	233	231	228

Topographic maps of small areas, such as hikers carry, are ideal for that purpose. Longitude is marked at the upper and lower borders, where the thin black North-and-South-running lines end. Some road maps, and all atlases, show longitudes—often very discreetly—near the top and bottom margins. Your public library probably has a collection of maps and atlases.

Longitudes are measured in degrees (°) and minutes ('). One degree, like one hour, is divided into sixty minutes. So 30' equal ½°, and 15' equal ¼°.

All longitude measurements start from Greenwich (near Lon-

Local Time of Noon

Jan	1–3	12:04	May	21–30	11:57	Sep	23–25	11:52
	4–5	12:05		31	11:58		26–28	11:51
	6–7	12:06	Jun	1–5	11:58		29–30	11:50
	8–10	12:07		6–10	11:59	Oct	1	11:50
	11–12	12:08		11–15	12:00		2–4	11:49
	13–15	12:09		16–20	12:01		5–8	11:48
	16–18	12:10		21–24	12:02		9–12	11:47
	19–21	12:11		25–29	12:03		13–16	11:46
	22–25	12:12		30	12:04		17–22	11:45
	26–31	12:13	Jul	1–4	12:04		23–31	11:44
Feb	1–22	12:14		5–11	12:05	Nov	1–14	11:44
	23–29	12:13		12–24	12:06		15–19	11:45
Mar	1–5	12:12		25–27	12:07		20–23	11:46
	6–9	12:11		28–31	12:06		24–26	11:47
	10–13	12:10	Aug	1–9	12:06		27–29	11:48
	14–16	12:09		10–15	12:05		30	11:49
	17–20	12:08		16–19	12:04	Dec	1–2	11:49
	21–23	12:07		20–23	12:03		3–4	11:50
	24–26	12:06		24–27	12:02		5–7	11:51
	27–30	12:05		28–30	12:01		8–9	11:52
	31	12:04		31	12:00		10–11	11:53
Apr	1–2	12:04	Sep	1–2	12:00		12–13	11:54
	3–5	12:03		3–5	11:59		14–15	11:55
	6–9	12:02		6–8	11:58		16–17	11:56
	10–13	12:01		9–11	11:57		18–19	11:57
	14–17	12:00		12–14	11:56		20–21	11:58
	18–22	11:59		15–17	11:55		22–24	11:59
	23–27	11:58		18–20	11:54		25–26	12:00
	28–30	11:57		21–22	11:53		27–28	12:01
May	1–7	11:57					29–30	12:02
	8–20	11:56					31	12:03

don). The numbers run westbound, through Europe west of Greenwich to the Americas. Eastbound they run through Europe east of Greenwich to Asia and Australia. Longitudes west of Greenwich are labelled "W", east of Greenwich "E".

If you are east of the standard time meridian of your zone, you must subtract a correction for the difference in longitude;

when you are west of the standard meridian you must add it to get local time of noon.

Here's a memory aid: The sun rises in the east and moves toward the west. So the sun will be south earlier at a place east of the standard meridian. Earlier East. To get the time of local noon subtract the correction from the Standard Time given in the table.

You have to Wait for noon West of the standard meridian. Add the correction.

Standard time zones are 15° of longitude wide and differ by one hour. From that follows:

1° longitude difference makes the correction 4 minutes.
15′ longitude (¼°) difference
makes the correction 1 minute.

To find South from the sun at noon you also need accurate time. You can get it anywhere in the world on one of several short-wave bands from stations WWV and WWVH. Or call for fifty cents 1-900-410-TIME from anywhere in the United States and hear it in plain English from the Naval Observatory.

Universal Time should not give you any trouble. Concentrate on the minutes. Any watch or clock will give you the hour in your zone.

Quartz wristwatches let us carry time confidently into the wilderness. Check yours by calling the time number about three days after you first set your watch.

Equal Altitude

There's another method for getting true direction from the sun, and it doesn't need accurate time. In fact it doesn't use a time-piece of any kind. All it takes is some of your time: You may have to wait, say an hour or more, to tell where North or South lies.

The basis for this method is simple. The height of the sun above the horizon after noon is the same as it was the same length of time before noon. You simply compare the length of an afternoon shadow with that of a morning shadow.

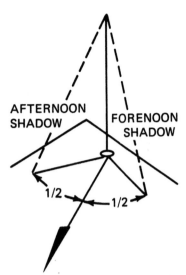

AFTERNOON SHADOW

FORENOON SHADOW

1/2 — 1/2

12.6 Finding North (South in Southern Hemisphere) from shadows of equal length before and after noon.

Start by marking the point where the shadow of the tip of a vertical object falls on a flat, level surface before noon. When the afternoon shadow is of the same length, make a second mark. To measure, all you need is a piece of string.

South lies on a line from the point halfway between the two marks to the base of the pole. Fold the string on itself to the halfway point.

The setup for getting South may vary from a column in a public square to a flagpole in a paved yard to a ski pole—point up— on a snow-covered veranda to a pencil mounted on a board lined with concentric circles.

Why not always use this method to avoid the mess of time corrections of the sun at local noon? You may not want to spend the time in the middle of the day. More to the point: The method isn't as practical at some times of the year in some areas.

Near the Arctic Circle in March and September the sun is so low the shadow may fall in the next county, if they have counties there. On the southern border of the NTZ, near the Tropic of Cancer, a pencil tip doesn't cast a shadow in June; the pencil swallows it.

Other places and times are less extreme. Try it yourself. (You don't have to worry about the sun's changing declination spoiling your experiment. It never changes more than 1/60 of a degree in one hour.)

South from Your Watch

Some people and books may tell you how to find South anywhere from your watch whenever the sun shines. Don't believe it.

They may try to convince you with what sounds like flawless logic: The Earth turns on its axis at a constant speed, once in twenty-four hours. The hour hand of your watch turns at a constant speed once in twelve hours. So it's obvious that a simple relation exists between the sun's position and the hour hand.

The relation is far from simple. For the watch method to work, the bearing of the sun should steadily increase by 15° per hour.

At the mid-latitude of our zone (45°N), in midwinter the hourly change in the six hours before and after noon varies between 10° and 15° per hour. In midsummer it varies there from 10° to 34°. At the southern edge of the zone, in midwinter the hourly change in the same six hours varies between 7° and 19°. In midsummer it varies between 0° and 4°.

That obviously can't work with the hour hand of a watch, which moves at a steady rate and does not vary with latitude and season.

You may wonder how sundials manage to show correct time. The part that casts the shadow, the gnomon, points to the celestial North Pole. For a dial with evenly spaced numbers to work, it must lie in the plane of the celestial equator, tilted at the proper angle with its highest point facing south.

In a museum you may see a portable sundial made before watches were available. Already in the 1500s there were models, the size of a pocket watch, with a hinge adjustable for the traveler's latitude. A built-in compass allowed you to orient it properly.

These sundials let you find the time from the sun. In the watch method for finding direction from the sun you are holding the watch horizontally, as you'd hold a compass. You can't hold it at the proper angle—high point due South—because you don't know yet where south lies.

Direction from a Moving Shadow

There is another method for finding direction that someone may try to sell you in print or on a dark night. Don't buy it.

You are supposed to find a flat, level surface and a vertical object with a sharp top—such as a flagstaff in a public square, or a ski pole, point up, stuck in the snow over a frozen lake. Add sunshine. Mark where the shadow of the tip falls. Make a second mark when the shadow has moved a bit.

The advocates of this method say you don't have to note the time, just connect the two points with a straight line that will point true East from the second mark and West from the first mark. If you look skeptical you'll be told "Everybody knows the sun moves from East to West, so its shadow will seem to move from West to East. What's hard to understand about that?''

Perhaps you seldom see snow on frozen lakes where you hike. Perhaps you often forget to take a public square with flagpole on your canoe trips.

Wouldn't it be nice if anywhere in this zone you could get directions whenever the sun shines, without correct time, calculation, or tables?

You can't and you have known why from the very beginning of this book. You were asked to stand facing south swinging your arm from the horizon left of you to the horizon through a point high up and due South. Later you found out that the arc you traced in the sky was the celestial equator.

That happens to be the track of Alnilam in the belt of Orion and the stars you met in Virgo and Aquila. They always rise due East, ride high when South of you, and set due West.

That is also the track of the sun around March 20 and September 22. On all other days the sun's track is parallel to this one. It's farther north in summer and farther south in winter, but the shape and high point are the same.

You can see that everywhere in this zone the sun's shadow moves not only from west to east but, at the same time, south— in the forenoon—and north—in the afternoon. In the first and last hours of sunlight the hourly movement of the shadow can be more south or north than east. So much for that general system designed for finding east.

You may say I could have skipped this system entirely, or simply mentioned that it doesn't work in this area. But that would have been unfair to the smart readers who remembered that around the time of noon the sun changes direction rapidly but height above the horizon very slowly.

Can you get a decent east or west around noon? Yes, within about forty minutes either side of local noon—roughly between 11:20 and 12:40 sundial time. At that time twenty or thirty minutes between marking the first and the second shadow of the tip will do it. Your guess of actual noon can be quite a bit off without getting outside the critical time period.

Don't count on finding the time of local noon by measuring the shortest shadow. At the critical time the shadow stays about the same length while its direction changes rapidly. That's why this method here works at that time of day and no other.

You can use the shadow of a monument cast on a public square, or a ski pole point up on a frozen lake, or a pencil somehow made to stand upright on a picnic table.

A line connecting the earlier and the later shadow marks will run west to east. The second mark will be about east of the first one.

The calculated maximum error near the Tropic of Cancer will be less than 4°, half that near the Arctic Circle.

In this area movement of the shadow of the tip of a vertical object may come in handy some day. Suppose you don't have the correct time, or don't know your precise longitude or the time of local noon for the day. Perhaps you have all the data but missed the time for finding South from the sun. Maybe a thick cloud hid the sun at the critical time.

13

The Arctic

Technically the arctic zone starts at the Arctic Circle near latitude 66½°N and ends at the North Pole, which is by definition at latitude 90°N.

More romantically it is known as the Land of the Midnight Sun. We could define it, quite accurately, as the part of the northern hemisphere where the sun at times stays above the horizon all day.

Perhaps you were taught in school that at the Arctic Circle on the night of the summer solstice, about June 21, the sun stays above the horizon even at midnight.

That was easy to teach and simple enough to have some hope of being remembered by pupils. If later you were lucky enough to take a North Cape Cruise around that time of year, you were in for a pleasant surprise. At the Arctic Circle part of the sun was visible at midnight for about two weeks before and after the solstice, from about June 7 to July 6. (All such dates may vary by a day or so, partly due to leap days.)

On or about these dates, the upper edge of the sun (upper limb is the technical term) just kisses your horizon. The sun appears to have just set when it starts rising.

The upper limb of the sun on the sea horizon, observed from sea level at standard air temperature and barometric pressure, is the definition of the moment of sunrise and sunset. In published predictions the effect of refraction—bending of the light rays in the atmosphere—is already allowed for. The prediction is just what you and I would call sunrise or sunset at the place and with the conditions mentioned.

But what happens around midnight just south of the Arctic Circle? Simple: Dusk runs into dawn. At such times it stays bright enough to read all night.

Earlier or later in the year—when the sun goes down for a while at the Arctic Circle—you may not be able to read. Farther south, the brightest stars will come out, but you can still make out where the sea and the sky meet.

That explains the "white nights" of Russian poets. Dusk runs into dawn in Leningrad (latitude 60°N) and other places near that latitude—Anchorage and Oslo, for example—for weeks on end.

For shorter periods you'll find the same conditions farther south, say near latitude 58°N in Juneau or Göteborg.

Even on a short cruise to Alaska or the North Cape, you'll be struck by the long dusk and the early dawns there. These extended twilights start long before you reach the Arctic. Visitors to Victoria, Canada—1250 miles (2,000 km) south of the Arctic Circle—remark on it.

It's all explained by our reference line, the celestial equator. Its high point, south of you, gets lower as you move north. The sun's path on any given day is virtually parallel to the celestial equator—higher in summer, lower in winter, but always parallel.

When the high point of the equator is high, the sun plunges below the horizon at sunset, giving you a short dusk. In Key West, it's dark before your second evening drink.

When the high point is rather low, the sun slides down at a shallow angle, giving you a long dusk.

At the Arctic Circle the sun's high point at the summer solstice is 47° above your horizon. At the winter solstice it is virtually on the horizon. The shortest day there lasts about three hours, with three hours dawn and three hours dusk.

Geography

Not to disappoint geography buffs, I add this:

The Arctic Circle, which separates the Arctic from the rest of the world, runs south of the Brooks Range of Alaska, through the north end of the Great Bear Lake of Canada, through the south tip of Greenland, passes just north of Iceland, then continues through the north tips of Norway, Sweden, and Finland.

It enters the Soviet Union's Kola Peninsula well south of Murmansk, near longitude 30°E. It then runs through the European USSR, crosses into the Asian USSR near longitude 65°E. After having passed through eleven time zones it leaves the Soviet Union, crossing the International Date Line in the Bering Strait, and returns to Alaska near longitude 169°W.

Basic Sky Facts

All your practical astronomy from the North Temperate Zone still works near the southern border of the Arctic.

You find the high point of the celestial equator, our reference line, due south and 90° minus your latitude above your horizon. On the Arctic Circle, in latitude 66½°, that makes 23½°.

You'll find few towns here designed with north–south streets or avenues. Magnetic compasses in the Arctic may act strangely due to the nearness of the magnetic pole. In some areas their readings vary between erratic and useless. Make local inquiries about an area's magnetic conditions. Local rangers, Mounties, boat captains, and pilots should know.

You may think the gods wanted to confuse travelers in the Arctic: Gyrocompasses also act up in arctic latitudes. They depend on the directive force caused by the rotation of the Earth around its axis.

That force is strongest on the equator and diminishes with latitude. It is zero at the pole. From latitude 70°N northward, the standard gyro cannot be trusted; at latitude 85°N it "probably becomes useless," according to U.S. Navy Hydrographic Office Publication No. 9.

You may think you don't need outside help to find South. Whenever it is dark anywhere in the Arctic you'll have Polaris to indicate North. If in doubt, check that you have the right star with the Big Dipper.

Polaris is virtually at the celestial North Pole. The celestial North Pole is as many degrees above your north horizon as you are north of the equator. There's the rub: In the northern part

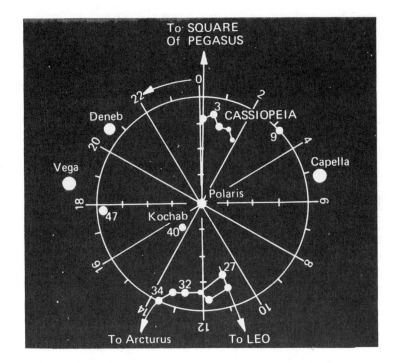

13.1   The northernmost stars.

of the Arctic, Polaris is so near the zenith that it's hard to tell where north lies.

On the Arctic Circle Polaris is 66½°, or three spans and three fingers. (You can check your measurements with Figure 2.2.)

If Polaris is low enough to find North from it, turn your back to it to face South.

If Polaris is too high, use your compass to find South. If you don't know the local correction, or when you are in an area where compasses are not to be trusted, you can get approximate South from the stars, close enough to place your strip map. When broken clouds make matching the map to the stars difficult call local star time to the rescue.

Appendix A gives rough star time for civil time for any day and year. Or look at the northern sky. Match the map of the Northern Cap (Figure 13.1) with the sky by turning the book. The number directly above Polaris is the star time of the moment, say 20ʰ.

Next, recall that local star time tells which hour of right ascen-

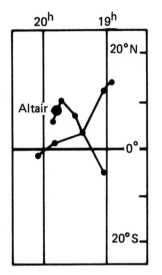

13.2 Aquila (Eagle) with first-magnitude Altair.

sion (RA) is due South of you. Say you get 20ʰ for star time now. Figure 13.2 shows a star, Altair, close to the 20ʰ line, and about 10° north of the equator.

If you are in latitude 70°N your equator is 20° (90 minus your latitude) above your south horizon. Altair will be about South and 10° higher, three hands above the horizon.

If you find Altair you may also find Deneb or Vega to clinch the identification. Better yet, Altair, all by itself (see Figure 5.8), is marked by two stars, one above and the other below.

Traveling North

When you travel north from the Arctic Circle, the celestial North Pole—conveniently marked by Polaris—will rise higher as your latitude increases. That will be obvious quickly if you travel by jet; by dogsled it'll take a little longer.

The Pole will rise 1°

for every 60 nautical miles of northing,

which is about 69 statute miles (111 km).

These figures are correct anywhere in the world. They are easier to remember when you know that one nautical mile was planned to equal one minute of latitude—1/60 of 1°—everywhere. The shape of the Earth causes a small change in the decimals. But on the average 60 nautical miles equal 60 minutes of latitude, or about 69 (statute) miles.

The metric equivalent is related to the fact that the meter originally was defined as the ten-millionth part of the distance from the pole to the equator, 90 degrees of latitude. Ten million meters are ten thousand kilometers.

When you divide 10,000 kilometers by 90 you get a bunch of ones. One degree of latitude then equals 111 km. The 111 km figure again is close enough anywhere in the world, but the shape of Earth changes the decimals with your latitude.

Anyway, the pole rises as you travel north. At the North Pole—by no coincidence—the celestial pole is directly above you.

The celestial North Pole, just like the one on Earth, is 90° from the equator. So as the pole rises when you travel north, the celestial equator drops 1° for every degree of latitude.

Fixed Stars

As the pole rises, more and more stars never set. They become circumpolar.

The three first-magnitude stars just north of the strip map (Figure 4.2) are already circumpolar at the Arctic Circle. They are Pollux of the Hexagon and Deneb and Vega of the single Triangle. As you go farther north, more and more stars become circumpolar. At the North Pole none of the stars north of the equator on the strip map ever sets.

The sinking of the high point of the equator means that some stars south of the equator on the strip will never rise. The stars in the southern gap have resigned already a few degrees south of the Arctic Circle.

At the Arctic Circle two first-magnitude stars disappear: Fomalhaut, the Loner, and Antares in the Double Triangle. With that loss the Double Triangle also becomes single.

As you approach the North Pole you'll slowly lose all the stars south of the equator on the strip map. The loss will be accelerated by extinction near the horizon: A star within 10° of the horizon appears one whole magnitude dimmer; within 4° two magnitudes. That's quite apart from frost smoke and other atmospheric conditions.

At the North Pole the equator lies in your horizon. Stars south of the equator never rise; stars north of the equator never set. The cap of northern stars seems suspended from the zenith, above the northern half of the strip.

13.3 Dome of sky at the North Pole. The equator of the strip falls in the horizon. Only stars north of the celestial equator are visible here, and they never set.

Sun, Moon, and Planets

You can figure out what happens to the sun when you watch it in the Arctic. It's easiest if you start watching from the North Pole.

The sun acts just like other stars. When it's north of the equator it never sets; when it's south of the equator it never rises. For example, when it's near Pollux it never sets, near Antares it never rises.

You'll find the path of the sun, the ecliptic, in Figure 8.2. There you can see that the sun is near Pollux in summer, near Antares in winter. You can also see that between some day in March and some day in September the sun is north of the equator. It will never set at the North Pole.

If you feed the right data into a computer it will calculate that the length of day there is about 190 days—from about March 19 to about September 25, or about 4,560^h hours.

The rest of the year the sun does not rise.

Near the Arctic Circle the shortest day lasts about three hours. At the pole for 175 days in a row the sun is continuously below the horizon. So it is not very meaningful to talk about an average length of day in the Arctic.

Also, you have to define very precisely what is above the horizon—upper limb, center, or lower limb?—and referred to which horizon—apparent, rational, or some other?

In the definition of twilight we used the apparent—visible—horizon. The time from sunset to when you can't read anymore is known as civil twilight. You don't have to define who reads what size print on what paper. Civil twilight starts when the center of the sun referred to the center of Earth is 90°50' from the zenith. It ends when that angle becomes 96°.

At that time it is far from dark. The twilight that follows is known as nautical twilight. It's used by ship's navigators morning and evening, weather permitting, to fix their position at sea (see Chapter 17).

In that light, bright stars and planets are already visible, and you can still make out the line that separates water and sky, the sea horizon. That's what you need for a marine sextant sight.

When in clear weather the sea horizon becomes vague, the navigator has to put his sextant away. But it's still not dark.

The remaining light is called astronomical twilight. It begins when the angle between zenith and the sun is 102°. Now the dimmer stars appear. That twilight ends when everybody would agree there isn't a trace of daylight left. The sun then is 108° from the zenith.

After six months of steady sunlight at the North Pole come unexpectedly about two weeks of continuous darkness. After October 9, you'll enjoy continuous civil twilights. Beginning October 25, navigational twilights every morning and evening cut into the civil twilights.

From November 14 to January 9, when the sun is far south, you will finally have times of twilight dim enough to qualify as astronomical. After that the twilights return to civil and navigational until February 17. Then come days of continuous civil twilight until March 19, give or take a day.

Then the sun rises and doesn't set for another six months.

The Planets and the Moon

The planets stay within a few degrees of the ecliptic. That makes approximate forecasting of their visibility in the Arctic quite practical.

For Mars, Jupiter, and Saturn you find the declination of each planet in the calendar in Appendix B. When the declination is marked N (for north) the planet may be visible all the way to the North Pole.

Declinations marked S (for south) must be less than 23° for planets ever to be visible at the Arctic Circle. If the declination is between 23°S and 0°, the planet may be visible in a part of the Arctic. Say the declination is 13°S. Subtract that from 90°. You get 77°. The planet may be seen as far north as latitude 77°N.

The same calculation works for fixed stars: You can find the declination of all first-magnitude stars in Appendix F. For example, how far north can you see Rigel? The table gives 8°S for Rigel's declination. Subtract that from 90°; you get 82°. You may see Rigel as far north as latitude 82°N.

Of course it must be dark or nearly so to see the stars or planets. So you won't see any on the North Pole from late March to late September.

The moon, like the planets, closely follows the ecliptic. Closely here means within, say, less than three fingers above or below it. To give a calendar of the moon's declination, which can change several degrees from one day to the next, isn't practical. You can't estimate the change from the moon's phase.

If it's dark and cloudless and you don't see the moon, it's below your horizon, unless Appendix B shows it to be within a day or so from new moon.

Enjoy the moon, and its light when it is visible.

At the North Pole

Let me take you back once more to the North Pole. Not because many readers will set up camp there to watch planets, stars, or the sun and moon.

You already know what conditions are at the northern edge of the North Temperate Zone, the Arctic Circle. By knowing what they are at the North Pole, you can imagine what they must be anywhere in between, anywhere in the Arctic.

In March, after several month of darkness and some weeks of twilight, the sun breaks the horizon. The first day only a sliver of the upper limb circles the sky. Then the sun spirals up, a little each day, exposing more of its face and finally clearing the horizon completely.

It continues to circle clear around the sky, a little higher each day. By the third week of June it reaches its highest point, 23½° above the horizon. It then spirals down a little each day.

By September its lower limb touches the horizon. That part disappears the next day. The sun keeps sinking day by day until even the upper limb has disappeared. Then a long twilight period begins. It ends when the sun is about 18° below the horizon, when by any standard it's dark.

You are now in the long winter, which is actually two weeks shorter than the summer.

Polaris is virtually overhead. The stars seem to circle the sky without rising or setting. They just run parallel to the horizon, which is at the celestial equator. That makes stars with north declination stay up all the time, stars with south declination never appear.

In the same manner, when the moon and planets are in north declination they never set. When they are in south declination they never appear.

That makes you lose sight of the moon for about half of every month. The periods when the moon is visible or not are unrelated to its phases and so can't be read from Appendix B.

But you can get the declination of Mars from Appendix B.

Jupiter is invisible for years on end. It rises and sets once in its trip around the sun, which lasts for twelve years. In the next years, it sets in October 1992 and rises again only in March 1999.

Saturn also rises and sets once in its orbit around the sun, which lasts 29½ years. Saturn last set in 1980 and will not rise again at the pole until March 1997.

Finding Directions in the Arctic

A Star to Steer By

When the stars are out you can keep a boat on course without watching the compass all the time.

When you are on the desired course, look for a star dead ahead, then steer for it. After a while, check the compass. If you are still on course continue toward the star.

You'll find a star not too close to the horizon but not very high up easiest to steer by. About a span above the horizon, but no higher than two spans, would be my choice.

You don't have to know the star's name or its constellation, as long as you can recognize it again after looking away for a moment. Find a star or two nearby stars to make a distinctive pattern with your chosen one. One above the other, or the left one of two at about the same altitude, will do nicely. A star or planet, brighter than all nearby stars, would be great. When no star is dead ahead, steer for the space between two stars.

That is like steering for a buoy. But this buoy is adrift. On your next compass check, you'll find the star has drifted off your course. For the next little while steer to the spot that's now dead ahead. When you steered before for the space halfway between two stars, now favor one star over the other.

When the original star or stars become too far off your course, check your compass and aim for another star or space between stars.

Note that I have not said the guiding star has drifted west. Only at the North Pole do all stars drift west. Elsewhere some stars—circling below the pole—drift east.

How fast do the stars or planets drift in the Arctic? Here's a rule of thumb for this zone: A star one to two spans above the horizon will drift about one finger (2°) in eight minutes.

You can aim for a star on foot, skis, snowshoes, or dogsleds. It's similar to what experienced hikers do. They pick a landmark by compass and move toward it. When they reach it, they pick a new landmark and repeat the process.

Polaris

For our purpose Polaris is always close enough—within half a finger—of the celestial North Pole. When you see Polaris you will know where North lies, and from that you can find South, East, and West.

At the Arctic Circle the pole is already 66½° (three spans and three fingers) above the horizon. For every 140 statute miles (220 km) you travel north, it rises another finger. Halfway to the North Pole, near latitude 78°N, it is almost four spans above the horizon. It becomes difficult to be quite sure in what direction Polaris lies from your zenith.

For that reason, and the increased distortion here, the Kochab–Polaris clock is not practical in most of the Arctic. Near the Arctic Circle, use Figure 12.4 to find star time.

Special Stars

In the southern half of the Arctic, the stars close to the celestial equator may indicate West at setting or East at rising, with errors of up to about 4° toward the South or North.

That may be good enough for some purposes, say for finding

or identifying a constellation. For checking the error of your compass, that is not good enough.

In theory, at the North Pole stars on the equator never rise or set. The three stars in Orion, Virgo, and Aquila—described and pictured in Figure 12.5—are not quite on the equator. That and effects in the atmosphere blur the moment of rising or setting beyond usefulness in the northern half of the Arctic.

Directions from the Sun

For finding direction, the rising or setting sun lets you down in the Arctic. For every degree the sun's declination changes, the bearing at sunrise and sunset changes by several degrees. Worse, much of the year the sun dawdles along the horizon. That makes the moment of true rise or set vague, and with it its direction.

The sun at noon gives you accurate south here regardless of latitude. But it needs correct time, and an often substantial correction for longitude, besides the one for the time of actual noon (Figure 12.6).

The sun at any time lets you find South in the Arctic. All you need is a watch.

The instructions for finding South are simple.

Hold the watch horizontally as you would a compass. Point the hour hand toward the sun. That's easy when the sun is close to the horizon. At other times, you'll find it more convenient to align the shadow of a toothpick, knife blade, or pine needle with the hour hand.

South will lie about halfway between the hour hand and the 12 o'clock mark on the watch dial. (See Figure 13.4.)

That works well enough if you only need rough direction, say, deciding which fork in a mapped road to take.

For more accurate work you must allow for daylight saving time, which in most countries makes your watch one hour fast. You must also allow for the difference in longitude between you and the reference meridian of your time zone. Unless it's only a few minutes, you should allow for the difference between apparent noon and 12 o'clock (see page 130).

How close to true South can you get with this method?

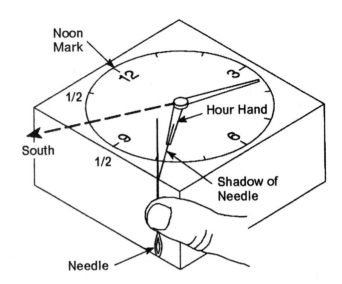

13.4 Hour hand for finding South. A clock or watch face lets you find South, halfway between the hour hand and the 12 o'clock mark. Hold a needle, as shown, so that it casts its shadow on the hour hand.

Assume you made the necessary corrections, aligned the hour hand perfectly with the sun, and halved the angle between it and the 12 mark exactly. Then the maximum error—in midsummer—on the Arctic Circle is a whopping 10°.

In latitude 72°, where sunrise and sunset bearings become vague, the error shrinks to 5°. Only near latitude 85° does the maximum error get down to 2°. You'd have to travel to within 1° of the North Pole to get the theoretical maximum error down to 1°.

The movement of a shadow cast by the sun in the Arctic can give you a fair East–West line.

You need a vertical object to cast the shadow, and a flat and level surface to receive it, such as a ski pole stuck, point up, in the snow on a lake.

The time to get East and West is around noon. That's when the sun changes direction most rapidly and altitude most slowly.

You don't need the time of local apparent noon to the minute. Any twenty- to thirty-minute period between forty minutes before or after actual noon will work.

Roughly estimate the time of noon by applying corrections for daylight saving time, longitude, and for date (from page 130)—unless the difference between apparent noon and 12 o'clock is only a few minutes.

Then mark the first point where the shadow of the tip falls, say half an hour before your estimated noon. Have lunch, or do whatever you like for the next twenty minutes or half hour. Then mark the second point where the shadow has moved to by now.

A line drawn between the two points runs roughly East–West. Continued past the second mark it points East; continued past the first mark it points West.

At the Arctic Circle that line will be within 2° from the true direction. The deviation from the true direction gets smaller as you approach the pole. But with the sun low, even at noon, the long shadows may create practical problems.

The equal altitude method you read about in Chapter 12 works in much of the Arctic when the sun is not too low at noon. But you'll need plenty of flat level ground and patience. Allow up to three times the height of your point for the shadow. You may have to start 1½ to 2 hours before noon to get accurate results.

14

The Tropics

The Tropics lie between the north and south temperate zones.

Perhaps you learned "between the Tropic of Cancer and the Tropic of Capricorn." That's correct. But were you told why these lines are called the Tropics?

The Greek root "trope" implies turning. What turns there is the sun. Around March 20 it is directly above the equator. It then moves north. Around June 21 it is directly overhead near latitude 23½°N. There it turns. It moves south and arrives again directly above the equator around September 22.

The sun then continues to go south, until around December 21 when it is directly overhead near latitude 23½°S. There it turns again, north toward the equator. Its estimated date of arrival is March 20.

A good defininition of the Tropics then is the area where the sun at times is directly overhead.

Cancer for the northern turning point and Capricorn for the southern are the constellations of the zodiac where the sun did the turning when these names were coined. The names are still used today. After all, you can't just change the labels on all maps every 2,000 years!

From north to south the Tropics are about 3,200 statute miles (5,200 km) wide, about the width of either of the temperate zones.

But their areas are very different. At the equator a degree of latitude and a degree of longitude are the same length. But all the longitude lines (meridians) meet at the poles, so the

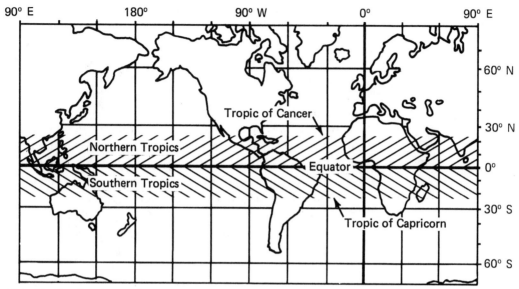

14.1   The Tropics, the area between the Tropic of Cancer and the Tropic of Capricorn.

greater the latitude the smaller the degree of longitude becomes.

At the equator 1° of longitude is about 69 statute miles (111 km). At the north and south borders of the Tropics that has changed to about 64 miles (103 km). At the poleward border of the temperate zones it has shrunk to about 27 miles (44 km). In human terms, it is like your waist measurement and your collar size.

You'll recall that the zones originally were zones of climate. The Tropics were the Torrid Zone. Indeed, in the lowlands they are on the warm side.

The climate of the northern and the southern Tropics could be described together. But naked-eye astronomy is very different in the two zones.

The Northern Tropics

The area between the Tropic of Cancer and the equator makes up the Northern Tropics.

This area has two periods each summer when the sun seems directly overhead for several days, once when the sun is on its way north, and once when it returns. Summer here—as everywhere in the northern hemisphere—means about the third week of March to the third week of September.

In the vicinity of the Tropic of Cancer these two periods, when at noon you stand on your own shadow, become one.

What strikes visitors from higher latitudes, say the NTZ, is the uniform length of tropical days. On either border of the Tropics the time from sunrise to sunset varies by less than three hours from the longest to the shortest day. Near the equator that difference is down to a few minutes.

Twilight is short everywhere in the Tropics. It's about twenty minutes from the brightest stars still visible to sunrise, or from sunset to the brightest stars becoming visible in the evening.

Apart from that and the periods of no shadows, the sky by day and by night continues the patterns of the North Temperate Zone.

At the Tropic of Cancer, the celestial equator is 66½° (90° minus your latitude) above your south horizon. So even in midwinter the sun at noon will be 43° above your south horizon.

The moon and the planets follow the sun's path within a few degrees, so they too will reach the highest point of their daily path roughly between halfway up the sky and directly above you.

To fit the strip map (Figure 4.3) of the stars to the sky you work here as farther north.

Face south. Measure 90° minus your latitude up from the point due South of you on the horizon. In the northern Tropics that will give, depending on your location, between 66° and 90°. Still facing South swing one arm, close to your body, first left, then back through the high point just found, to the right. That operation sweeps the celestial equator.

Next imagine the equator on the strip map on top of the one you have just drawn. The line that's marked with local star time should be due South of you, running straight down from the high point toward the horizon.

If you recognize a superconstellation, a constellation, an asterism, a star of planet near that line, align the star map to match the sky. If you have trouble getting a match, perhaps on a partly cloudy night or when you haven't watched the night sky for months, get the star time on the meridian from Appendix A or calculate it yourself as shown in Chapter 3.

Matching the star map of the northern sky will be easy. To find the spot for the center of the map, face North and measure your latitude upward from your horizon. Near the northern edge of the area that'll be about a span; near the southern edge the center will be on your horizon.

Now turn the circular star map until a constellation you recognize matches the sky. If you already know your star time, turn the star map until the RA on top of the polar map matches the one south of you on the strip map. (See Figure 14.2.)

In the northernmost part of the Tropics you can use Figure 9.2 to identify the brightest stars and the stars attached to the southern margin of the strip map. They'll be mainly Canopus, Archernar, Alpha and Beta Centauri, and the constellation Crux, the Southern Cross.

But as you get nearer to the equator you'll find the map of the stars around the celestial South Pole handier (Figure 14.3).

To match that map to the sky here (and only here) imagine where the celestial South Pole should be: as many degrees below your South Point on the horizon as you are north of the equator. For example, in latitude 15°N it would be 15° below your south horizon.

That locates the center of the map. The current star time—the star time you found earlier for the strip map—must be on top to match the sky. Or work it the other way. Match the south polar map to the stars you see and use the RA now on top to adjust the strip chart.

14.2 *Dome of the sky at the northern border of the tropics: North cap, strip map, and south cap placed at the Tropic of Cancer.*

Finding Directions
in the Northern Tropics

A Star to Steer By

When the stars are out, you can steer a boat here too without staring all the time at the compass. Or you can walk watching the trail ahead rather than a compass needle.

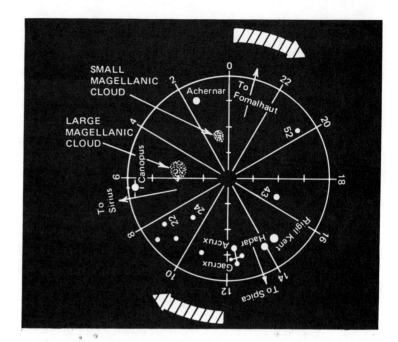

14.3   The southernmost stars.

The technique is simple. You put your boat or yourself on the wanted course by compass. Then you find a star straight ahead, and not too high up. A span or so above the horizon is most relaxing. You don't have to know the name of the star or its constellation, as long as you can find it again.

After a little while you check that you are still on course, and that the star is still leading you in the right direction. When it has drifted too far off, hold a little to one side of it. Eventually you'll have to find another star, perhaps one a little higher or lower.

Polaris

In the Tropics Polaris is less and less help for finding North the farther South you are. Observed from only a few degrees south of the Tropic of Cancer, Polaris becomes a third-magnitude star due to extinction near the horizon. Also the Big Dipper, which elsewhere helps you find or identify Polaris, disappears here for weeks on end.

When you are halfway or more into the Northern Tropics, don't look for help from Polaris in finding North.

Polaris is hard to spot below 10°N. Take the word of Captain Slocum and the many navigators who followed him. They always reported when Polaris finally reappeared after weeks and months below the horizon. It was usually north of Panama, 9°N.

As for a star near the south celestial pole, forget it. You aren't even supposed to see it until you get to the equator.

Equatorial Stars

But there's a method of getting East and West from the equatorial stars. Here they are especially good signposts.

Stars directly on the celestial equator rise due East and set due West everywhere. Here they rise and set steeply. That makes them change bearing very slowly after rising and before setting.

At the northern edge of this zone they take 10 minutes to change bearing by a single degree. In latitude 5°N they take 46 minutes to do so. No hurry here, and it doesn't matter if the horizon is not exactly at your level or free of obstructions.

No first-magnitude stars are close to the celestial equator. But three constellations that you can identify by such stars contain stars almost on the equator. They are Orion, identified by Betelgeuse and Rigel; Virgo marked by Spica; and Aquila found from Altair (Figure 12.5).

In Orion your guide is a star in the belt, the asterism formed by three stars almost in line, and about halfway between the two first-magnitude stars. I strongly advise you to use the middle star, epsilon. It's about a half-finger from the equator, but it has one great advantage: It remains the middle star wherever you see it.

Delta is virtually on the equator and would get you 1° closer to true East and West. But in the southern hemisphere it seems to trade places with the third belt star. If you used the wrong end of the belt, you'd be a whole finger off.

In Virgo you have a choice of two stars both very close to the celestial equator. One (zeta, the dimmer one) is about one hand from Spica in the direction of Arcturus. You'll find the other one (gamma) when it's dark enough to see the whole constellation. In Figure 5.8 it is where the virgin's head meets her torso.

In Aquila the star (theta) nearest the equator is a hand from Altair in the single Triangle. Altair has two outriders that point toward theta in the direction of the dimmer of the two stars.

Everywhere it is easier to recognize a setting star that you may have watched for hours than one just coming up. But here you have time to spare for positive identification before the rising star changes bearing.

The following table gives approximate star time of rise and set of the guide stars. With the longest nights you normally have a couple of risings or settings to get East or West from these stars.

Local Star Time[h]	Constellation	Phenomenon
2¼	Aquila	theta sets
6¾	Virgo	gamma rises
7½	Virgo	zeta rises
11½	Orion	delta and epsilon set
14¼	Aquila	theta rises
18¾	Virgo	gamma sets
19½	Virgo	zeta sets
23½	Orion	delta and epsilon rise

South from Crux

At times the Southern Cross can point south for you. The two stars that form the long arm of the Cross, Gacrux and Acrux, have almost the same RA. So when the cross is upright it points almost to the celestial South Pole, which is below the South Point on your horizon.

Star time then is $12\frac{1}{2}^h$. This trick isn't accurate enough to adjust your compass, but it tells you which fork in the road to take. For such decisions the Cross does not have to be exactly upright either. (Don't worry that the Cross may be upside down. It never will be here.)

Direction from the Sun

Instead of using the rising or setting of a second-magnitude star, we can use sunrise and sunset to find directions. No chance of getting the wrong star, no waiting up and finally missing it anyway! The sun is unmistakeable, and gives fair warning when it's about to rise and set.

Only at the equinoxes—about March 20 and September 22— does the sun rise and set everywhere due East and West. At all other dates the direction of rising and setting depends on how far north or south of the equator the sun is on a given day (its declination) and your latitude.

Here your latitude has very little effect on the bearing of the sun when it rises or sets. On a given day the bearing is virtually the same all through the Tropics. That makes sunrise and sunset excellent times for finding a true bearing here.

These bearings remain within 1° for at least 10 minutes before sunset and after sunrise, everywhere in the Northern Tropics.

Of course the usefulness of the sun for finding South at noon doesn't change radically when you cross the magic Tropic of Cancer.

When the sun at noon is overhead, or nearly so, it's no use for finding South. But far from the equator, and especially in the months from November through February, the noon sun is useful.

As in the temperate zone you'll need accurate time and two corrections: one for the real sun passing south of you up to a quarter of an hour early or late, the other for local time differing from the zone time your watch is set for. For that you need to know your longitude. Don't forget to take off one hour for clocks being fast when daylight saving time is being used.

It may be more trouble than it's worth to you. But if you want to work it out, use the table on page 130.

The equal-altitude method of direction finding shown in Figure 12.6 works here when the sun is not too high. When it is, the short shadows make comparison difficult.

At such times you can try getting direction by this method from the moon. The moon's declination may be far enough

---

Bearing of Sunrise in the Tropics (in degrees)

Date		Latitude, North or South 0°	23½°	Date		Latitude, North or South 0°	23½°
Jan	1	113	115	Jul	1	67	65
	16	111	113		19	69	67
	25	109	111		28	71	69
Feb	2	107	109	Aug	5	73	71
	8	105	106		12	75	74
	15	103	104		19	77	76
	20	101	102		25	79	78
	26	99	100		30	81	80
Mar	3	97	98	Sep	5	83	82
	8	95	95		10	85	85
	13	93	93		16	87	87
	18	91	91		21	89	89
	23	89	89		26	91	91
	29	87	87	Oct	1	93	93
Apr	3	85	85		6	95	95
	8	83	82		11	97	98
	13	81	80		17	99	100
	19	79	78		22	101	102
	25	77	76		28	103	104
May	1	75	74	Nov	3	105	106
	8	73	71		10	107	109
	16	71	69		18	109	111
	26	69	67		27	111	113
Jun	10	67	65	Dec	11	113	115
	30	67	65		31	113	115

---

south to give you shadows of decent lengths. Figure 10.2 gives approximate times of the moon's southing at the main phases. Add a little less than one hour for each day after every phase.

You can calculate in your head the times when the moon is highest in the sky. The full moon is opposite the sun, so it is highest at midnight. (You can get the dates of full moons from many calendars or in Appendix B.)

Subtract about one hour for each day before the moon is full; add about one hour for each day after. For example, if the moon is full on the 15th, from the 13th to the 17th the times

Bearing of Sunset in the Tropics (in degrees)

Date		Latitude, North or South		Date		Latitude, North or South	
		0°	23½°			0°	23½°
Jan	1	247	245	Jul	1	293	295
	16	249	247		19	291	293
	25	251	249		28	289	291
Feb	2	253	251	Aug	5	287	289
	8	255	254		12	285	286
	15	257	256		19	283	284
	20	259	258		25	281	282
	26	261	260		30	279	280
Mar	3	263	262	Sep	5	277	278
	8	265	265		10	275	275
	13	267	267		16	273	273
	18	269	269		21	271	271
	23	271	271		26	269	269
	29	273	273	Oct	1	267	267
Apr	3	275	275		6	265	265
	8	277	278		11	263	262
	13	279	280		17	261	260
	19	281	282		22	259	258
	25	283	284		28	257	256
May	1	285	286	Nov	3	255	254
	8	287	289		10	253	251
	16	289	291		18	251	249
	26	291	293		27	249	247
Jun	10	293	295	Dec	11	247	245
	30	293	295		31	247	245

for your observation will be, approximately, the hours given below:

13th	14th	15th	16th	17th
10 PM	11 PM	Midnight	1 AM	2 AM

Some people suggest using the hour hand of your watch to find South. That works near the North Pole. Here it is useless. Others suggest using the westward movement of a shadow to find East. That is generally unreliable everywhere, except near noon.

I can see one application here. Mark where the shadow of the tip of a vertical pole falls onto a flat, level surface at sunrise. Mark it again at sunset. Connect the two points. You have a good West–East line, with West where the morning shadow ended.

That will work accurately where the horizon is unobstructed, at sunrise and sunset, say on a prairie. It works because on any given day the sun rises and sets virtually the same distance from the point north—or south—of you. You can, of course, start by marking the sunset shadow, get a night's sleep, and mark the sunrise in the morning.

On the Equator

The equator, halfway between the North Pole and the South Pole, divides the southern from northern Tropics and the southern hemisphere from the northern.

The equator enters South America from the Galapagos Islands, runs through northernmost Ecuador, the southern corner of Colombia, and northernmost Brazil.

It enters Africa through Gabon, continues through Congo, Zaire, and southernmost Uganda, crosses Kenya and the southern corner of Somalia.

It crosses the Indian Ocean and runs through the islands of Sumatra, Borneo, and Celebes and a few smaller ones on its way to the Galapagos and South America.

Nobody lives on the equator. Sailors call it "the line," and a line it is. As you may have learned in first-year geometry, a line has no width. Nobody, not even an ant can live on a line.

So why do I give space to the sky over the equator? Because some rather unexpected astronomy happens there. The celestial equator is directly above every point of the equator on Earth. That's certainly not unexpected. But the celestial North Pole and South Pole are 90° from the celestial equator. That puts them on your horizon wherever you are on the equator.

The equator is the only place on the surface of Earth where you can see both poles at once.

Figure 14.4 shows how our strip map here forms a vertical

arch over you. It also shows both the northern and southern polar caps, each half above, half below your horizon. That simple arrangement has several effects.

- On the equator all stars set. There are no circumpolar stars or constellations.

- As if to make up for that loss, only on the equator can you see all stars. You can see all the constellations, 88 by official count. You may have to wait a few months for a certain constellation to be above the horizon during hours of darkness. But in the course of a year you'll see them all—from Ursa Minor near the North Pole, to Octans near the South Pole, and every one in between.

- A body on the celestial equator here passes directly above you when it is highest in the sky. The sun is on the celestial equator at the equinoxes, in March and September. So on the equator the sun is highest in the sky at these times. Not in their midsummer as in both temperate zones, the Arctic, and the Antarctic.

- Stars on the celestial equator pass directly above you every day, just as the sun does at the equinoxes. They rise—as everywhere else—due East of you. They set—as everywhere else— due West. Whenever you see one of the stars on, or almost on, the celestial equator it will be east until it passes overhead; then it'll instantly change to being due West of you until it sets.

  The stars that everywhere give you East and West at rise and set on the equator are compasses virtually all night. These stars in the belt of Orion, in Aquila, and in Virgo only fail you when they are close to being directly overhead. Then it's hard to tell on which side of the zenith they are. At such times simply use another one of these stars. There will usually be at least one other to guide you.

- As you travel from higher latitudes—whether north or south— to lower ones, the celestial equator rises. Put another way, the horizontal center line of our strip map, matched to the sky, will make a steeper and steeper angle with your horizon.

  On the equator the angle is 90°. It can't get any larger. What happens then? Plenty. Please read on.

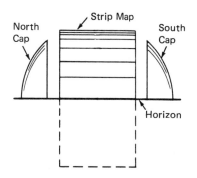

14.4   Dome of the sky at the equator: North cap, strip map, and south cap placed at latitude 0°.

The Southern Tropics

The Southern Tropics start at the equator and end at the Tropic of Capricorn.

The great surprise for most people as they travel south from the equator is this: The sun is north of you at noon most of the year.

The only time it will be south of you comes after it has been directly overhead and still keeps going south. It can't be more than 90° above your north horizon. When it goes 1° farther up, we call that 89° above the south horizon; two degrees farther we call 88°; and so on.

After the sun has reached the Tropic of Capricorn and returns, the numbers will go up again until they reach 88°, 89°, and finally 90°, directly above you again. Then it will be north of you until it gets again directly above you, the next time.

The sun still rises here on some easterly bearing and sets somewhere near West.

That, by the way, explains how long and short days, summer and winter, are reversed in the southern hemisphere. From about March 20 to September 22 the sun rises everywhere north of East, sets north of West.

In the northern hemisphere, where the sun moves by way of South, that makes for long summer days. In the southern hemisphere the sun moves by way of North. That makes for short winter days in the same six months, say April through most of September.

The long and short days don't create summer and winter. They just reinforce them. The main cause of the season is the high or low angle of the sun. It's summer when the sun is over your hemisphere, and high in the sky at noon. It's winter when the sun is over the other hemisphere, and low in the sky at noon.

Since the planets hug the annual path of the sun, it's no surprise that they too will be north of you much of the time.

The moon too follows the path of the sun within a few degrees. That places the moon in your northern sky most of the time. At times it will seem right side up, at others upside down.

You don't need a telescope to find which side is up. Just remember one or two prominent features. For example, in the

14.5   The full moon seen north of the equator (left) and south of the equator (right).

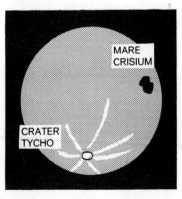

MARE CRISIUM

CRATER TYCHO

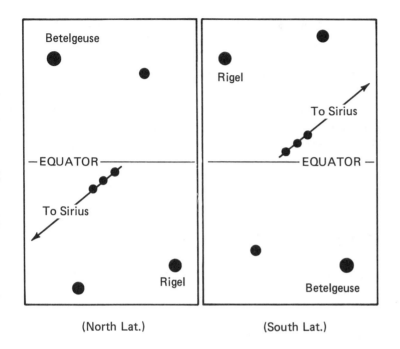

14.6 Orion seen north of the equator (left) and south of the equator (right).

(North Lat.)   (South Lat.)

flat light of a full moon the "rays" of the crater Tycho stand out. They radiate from that crater like the meridians from either pole on a globe.

One ray points roughly at an oval dark area near the edge of the moon's disk, the Mare Crisium (the Sea of Crises). Here the moon looks upside down, as in a telescope in the north.

Familiar constellations may look upside down south of the equator.

Orion, which straddles the celestial equator, is very prone to do so when the top edge of our strip map becomes the bottom edge in the southern hemisphere. But you have to look closely, or know Orion well, to notice the switch. When you see it in south latitudes above your northern horizon it looks like the right part of Figure 14.6. In north latitudes, above your southern horizon, it looks like the left part.

When you see the constellation, look for Rigel, the bluish first-magnitude star. Its name in Arabic means "foot." This foot here is oddly placed—at the top.

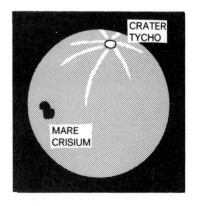

Another clue is reddish Betelgeuse, the other first-magnitude star in the constellation. It is one of the shoulders of Orion and switches places with the foot south of the equator.

The best signpost is the arrow pointing to Sirius. The three stars that form Orion's belt seem to point to that brightest of fixed stars. They point down to the left in the Northern Tropics; they point up to the right after you have crossed the equator. Just look for bright Sirius, a little more than a span from the nearest belt star.

Finding Directions in the Southern Tropics

The methods that work in the Northern Tropics—aiming at a star, and equatorial stars at rise and set—work here just as well. So does the sunset bearing combined with a sunrise bearing. The methods found useless—the hour hand of a watch, or the movement of a shadow—are no better here.

Other methods—like the sun at noon or equal lengths of shadows—work some of the time, depending on the height of the sun at noon where you are on a given day.

One method works beautifully here: the bearing of the rising or setting sun. Two factors combine to make it so good here.

1. On any given day the sun rises or sets virtually on the same bearing everywhere in the Tropics, from the Tropic of Capricorn to the equator and on to the Tropic of Cancer. (See tables on pages 158 and 159.)

2. Nowhere in the Tropics does the bearing just after sunrise or before sunset change by 1° in less than ten minutes. If a building or a hill obstructs your view of the true horizon, you still get a good bearing.

These two facts may have helped the Polynesians visit and revisit the islands of the Pacific. Their language and customs, from New Zealand to Easter Island and from Tahiti to Hawaii, bear witness to their achievements as navigators.

Can a bearing at sunrise and one at sunset let you reach a distant island and find your way back home again? Yes.

I once owned a small schooner built in Nova Scotia. When the sails were trimmed right, she would sail herself on any course without anyone steering. And she'd do that all day, or all night. The time of no change in bearing at sunrise and sunset was certainly long enough, even at the edge of the Tropics, to trim and fine tune sails and tiller.

We don't know that the Polynesians had discovered the secret of building sailboats that steer themselves. It is likely, though. If one boat of the many they built happened to steer herself, word would soon spread. Everyone would try to build such boats, not just for long voyages, but also for fishing and hauling traps.

But even without self-steering, there's a way offshore to keep a course for hours on end: steering by the swells. These long waves roll on, even while the local breeze changes direction.

To steer by the swells you put your craft on the course based on the bearing of sunrise or sunset. You observe how the swells run in relation to your craft, say at sunrise. Then you and the helmsmen who relieve you try to keep her at the same angle to the swells all day. At sunset you correct the course and steer the boat all night by the new angle of the swells.

Can that be done? Captain Voss set out from Victoria, British Colombia, in 1901 to sail around the world in a boat smaller than Captain Slocum's. Much smaller. *Tilikum's* hull was an Indian dugout, 5½ feet (165 cm) wide.

An unexpected green wave washed his helmsman overboard, and with him disappeared the ship's compass. No spare aboard.

Voss was then about 600 nautical miles southwest of Suva, Fiji Islands, and 1,200 miles from Sydney, Australia. He still had the tools needed for navigation—octant, chronometer, tables, and charts—and a lifetime's experience in using them.

> The guides I had to steer by were the sun, moon, stars and ocean swell, but I soon discovered that the ocean swell was by far the best to keep the boat making a good course.
>
> Then, again, I was obliged to use the heavenly

bodies to get the set of the swell. The only trouble I
had in finding the course was when after I got up
from sleep and found the weather thick and overcast.

But how could the Polynesians get the direction of the swell?
You already know. From the bearing of the rising or setting
sun, which is practically the same everywhere in their islands.

That lets you make a portable sun compass. At home you
set out a board with something—a dry thorn perhaps—in the
center to cast a shadow. Mark the shadows every few days,
morning and evening, and label the marks with the dates.

If you want to sail from Tahiti to Hawaii, ask an old-timer
for the course. He'll put it on your board, close to what we
call North. The true course is a bit west of North. But he will
want you to make your landfall upwind of the destination. That
way you'll slide downwind at the end of the voyage, guided
by a mountain peak.

At the beginning of the voyage, attach the sun compass to
your vessel with the course mark pointing to the bow. When
the shadow at or shortly after sunrise, or before sunset, falls
on the mark for that day, you are on course. Note where the
swells hit your craft and keep them there until the next check.

My ideas on Polynesian navigation were first published in
Sky and Telescope in June 1967. Since then Mr. David Lewis
sailed his catamaran *Rehu Mona* from Tahiti to New Zealand
by the swells, without compass or other navigational instruments, as reported in *National Geographic* in December 1974.

15

The South Temperate Zone

The South Temperate Zone (STZ) stretches from the Tropics to the Antarctic. More precisely, from the Tropic of Capricorn to the Antarctic Circle, from about 23½°S to about 66½°S. From the north to south borders that is about 3,000 statute miles (4,800 km). That's a little less than half the distance from the equator to the South Pole.

As you'd expect in a temperate zone, the length of day varies here more than in the torrid zones, but less than in the frigid zones.

15.1   *The South Temperate Zone, the area between the Tropic of Capricorn and the Antarctic Circle.*

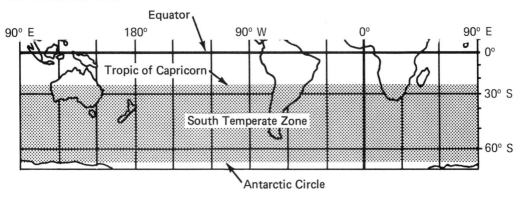

Wherever you live in the South Temperate Zone you'll be aware of the length of day, the time from sunrise to sunset, changing with the season. At the northern boundary of the STZ, it already varies from 10¾ hours in midwinter to 13½ hours in midsummer, a difference of close to three hours. (Midwinter, you'll recall, here falls in the second half of June, midsummer around Christmas.)

In the center of the zone, near latitude 45°S, the length of day varies from 8¾ hours in midwinter to 15½ hours in midsummer. That's a difference of close to seven hours.

At the south border of the STZ, near latitude 66½°S, the midwinter rise and set times are uninteresting since the sun doesn't even come one half-finger above the horizon. Just across the boundary, in midsummer the sun never sets, Midnight Sun. At noon on the same days it is 47° above the horizon.

More typical of the sun's angle above the horizon at noon is the middle of the zone. In latitude 45°S the angle varies from 21½° in midwinter to 68½° in midsummer. About the same as in Portland, Oregon; Burlington, Vermont; and Milan, Italy.

Geography

By far the largest part of South America is in the Tropics. But Chile, Argentina, and Brazil—except their northern corners—are in the STZ. So is much of Paraguay, and all of Uruguay.

In Africa the Tropic of Capricorn runs through Walvis Bay, Kruger National Park, the southern tips of Mozambique, and the island of Madagascar. Any land south of that line is in the STZ.

The Tropic of Capricorn enters Australia near its westernmost point and leaves it near Rockhampton on the east coast. The land south of that line and all of New Zealand are in the South Temperate Zone.

The STZ ends at the Antarctic Circle, near 66½°S latitude. All three southern continents—South America, Africa, and Australia—miss it by wide margins.

South America with Cape Horn, near its tip in latitude 56°S, comes closest—short about 700 miles (1,100 km).

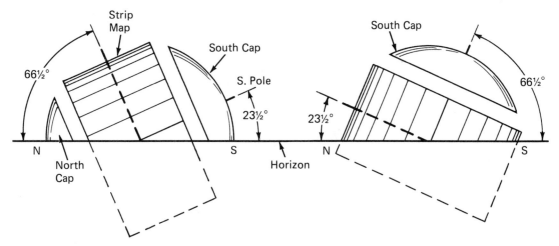

15.2   Dome of sky in the STZ: the north cap, strip map, and south cap. Left: in place at the Tropic of Capricorn, the northern border of the South Temperate Zone. Right: in place at the Antarctic Circle, its southern border.

Australia with Wilson's Promontory, a bit south of Melbourne, in latitude 39°S comes next—short about 1,900 miles (3,000 km).

Africa with the Cape of Good Hope near latitude 34°S—misses reaching the Antarctic Circle by about 2,200 miles (3,700 km).

The shortness of the southern continents makes it easy to give the geography of the southern border of the STZ. It crosses the Atlantic, Indian, and Pacific oceans in latitude 66½°S. (You'll read in the next chapter about the only sliver of land it crosses.)

The southern continents stretch over 56, 39, and 34 degrees from the equator. Contrast that with 78, 71, and 71 degrees for the northern continents—North America, Europe and Asia.

Fixed Stars in the STZ

Unless you have watched the sky in the Southern Tropics you'll be unprepared for what's profoundly different in the southern hemisphere.

1. The stars—and sun, moon, and planets—rise in the east, but here they reach their highest point north of you before setting in the west.

2. Constellations face the same direction as in the northern hemisphere, but up and down are reversed.

   Here Acrux is on top of the Southern Cross when Crux is highest in the sky. Gacrux, the dimmer of the two stars, is at the foot. North of the equator it's the other way around.

The Southernmost Stars

Find your latitude from a map or atlas, or by asking a scout or a local libararian.

The celestial South Pole here is as many degrees above your southern horizon as you are south of the equator.

For example, at the northern limit of the South Temperate Zone, in latitude 23.5°S, the celestial South Pole is 23.5° above the point due South of you on the horizon.

Face South, using a compass adjusted for the local correction, called declination on land, variation at sea.

Measure the distance from your horizon to the South Pole— here one span and two fingers.

Remember the spot just found, possibly by two or three nearby stars. It is the local center of the southern cap, in which the stars rotate clockwise around the celestial South Pole.

If you have watched the night sky farther north—in the Tropics or the North Temperate Zone—you'll find many familiar stars and constellations here. The nearer to this zone you have been star watching, the more friends you'll rediscover.

You may have been watching in the southernmost North Temperate Zone where at times the brightest southern stars rose a little above the horizon. Then you'll recognize some of the key players here.

The most prominent star in Figure 15.3 is Canopus near star time 6½ʰ. It is the second brightest (magnitude -0.9) fixed star after Sirius (magnitude -1.6).

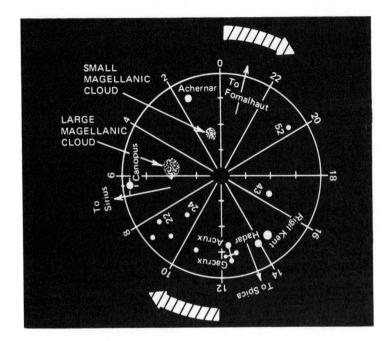

15.3   The southernmost stars.

You could have seen Canopus already from South Carolina, about two spans south of Sirius when they are both highest in the sky.

Achenar, a first-magnitude star, can be seen already in central Florida near star time 1½^h.

The constellation Crux, the Southern Cross, becomes visible near star time 12½^h already in Key West.

Next door to Crux you'll find the two first-magnitude stars, Hadar and Rigil Kent, often called Alpha and Beta Centauri. They are only four fingers apart and follow the track of the stars that form the short arm of the Cross. They reach their highest point in the sky near star time 14½^h.

Fomalhaut, the lone first-magnitude star not in a supercon-stellation in Figure 15.4 is near RA 23^h.

Antares, near RA 16½^h—a little outside the map area—is the brightest star in Scorpius.

If you recognize the Scorpion, the Southern Cross, or Canopus from its superior brightness—planets never get near it—you can get the approximate star time. Just orient the star map by turning it until it matches the sky. The numeral on top will give you local star time.

One star you will not find in Figure 15.3 is a star to mark the celestial South Pole. There is no helpful Polaris in the southern sky.

A rough way to find the pole is this: First find the Southern Cross. Then measure carefully the distance between the two stars that form the long arm of the cross. The South Pole lies in the direction of the brighter star (Acrux) and 4½ times that distance away.

The line from Gacrux through Acrux misses the pole by about two fingers. For greater accuracy do this: On that line, at the point found by measuring 4½ times the distance, measure, two fingers at a right angle to the line and on the side of Hadar and Rigil Kent. (See Figure 16.4.)

Don't worry about confusing Crux, the Southern Cross, and the False Cross. That asterism straddling the 9ʰ RA line has four second-magnitude stars. Crux has three brighter ones (first magnitude or almost so) and one much dimmer one (in the short arm of the cross). Alpha and Beta Centauri follow Crux by less than two hours. They are about five hours astern of the False Cross.

The star at the longer end of the long arm of the False Cross is navigational star #22 or Avior. Its Bayer name is Epsilon Carinae, carina meaning "keel."

This constellation is a result of the breakup of the constellation Argo, the Ship. Other parts are Puppis (the stern), Vela (the sails), and Pyxis (the compass). Canopus also is in Carina, as is another navigational star, #24 Miaplacidus (magnitude 1.8).

None of the constellations that once made up the Ship are easy to make out. But the constellation to which Achernar belongs, Eridanus the River, seems hopeless to me. It meanders between RA 1ʰ and 5ʰ, from the equator, near Rigel, to declination 60°S with only that one bright star in the whole riverbed.

On the other side, trailing near 16ʰ Rigil Kent and somewhat closer to the South Pole, is a constellation aptly named Triangulum Australe, Southern Triangle. Its sides are almost the same length. Its brightest star, alpha Trianguli (magnitude 1.9), is navigational star #43, stage name Atria.

Near RA 20½ʰ is navigational star #52, Pavo (Peacock), magnitude 2.1, in a constellaiton with the same name.

At RA 22ʰ is Al Na'ir, navigational star #55 (magnitude 2.2) in the constellation Grus, the Crane. This star seems to lie about halfway between Pavo and Fomalhaut.

The southern polar cap is a challenge. Its constellations are hard to make out. Third- to fifth-magnitude stars are needed to come up with bare outlines. Most of them take more effort than even northern constellations to connect with their names.

Stars in the southern cap never set if their south declination equals at least 90° minus your south latitude. For example, in latitude 23.5°S stars with declination 66.5°S or greater always stay above your horizon. They are said to be circumpolar there.

At that northern edge of the South Temperate Zone none of the brightest southern stars that you may already have seen farther north qualify. No first-magnitude stars are that near to the celestial South Pole.

To see the Southern Cross at all times of every night you'd have to be south of latitude 33°S.

There the two nearest galaxies—the Large and the Small Magellanic Clouds—are also visible all night, every night.

You may have noticed, even from my sketchy description, that the half of the cap between RA 6ʰ and 18ʰ is much more populated by bright stars than the other half, as if most toys in Santa's bag had drifted to the bottom.

Here is a more scientific explanation: The Milky Way (Figure 11.1) lies near the bottom of this map; Crux and Centaurus are right in it. And chances for finding bright stars are far greater near the flattened disk of our galaxy than away from it.

Of course there's more to your night sky than just the area near the celestial South Pole.

The Southern Strip Map

The technique for tracing the celestial equator in the STZ is about the same as in the other zones.

The celestial equator touches the East and West points on

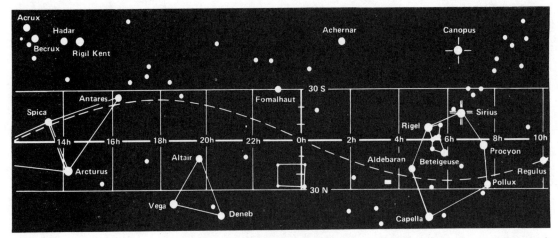

15.4   *Star map for the southern hemisphere, including all second-magnitude stars. This map can be used at any star time. Stars and hour lines appear twice on the map. To align map and sky use the hour nearest the fold. The planets Mars, Jupiter, and Saturn will never be far from the dashed curve, the ecliptic.*

your horizon. Here it rises to its highest point at 90° minus your latitude degrees above the North Point of your horizon.

For example, in latitude 23.5°S, the high point is 66.5° above your North Point. Get your latitude from the left or right margin of a map of your area, or ask an expert, perhaps a librarian or a pilot. After dark, face true North using a compass corrected for the local error. If you started tonight's star watching by facing the celestial South Pole, you only have to turn clear around.

Measure the distance from your horizon to the highest point. In the example that's three spans and three fingers. Point to it with your arm outstretched, and remember that spot.

Next, swing your arm, close to your body, left to the horizon, back to the high point, and then right to the horizon. As you faced North, you pointed to the West and East points. Your hand, going through the high point, traced the equator in the sky.

In a moment you'll align the equator on the strip map with the equator in the sky you just traced. That equator circles the sky, just as the equator on Earth circles the globe. At any given

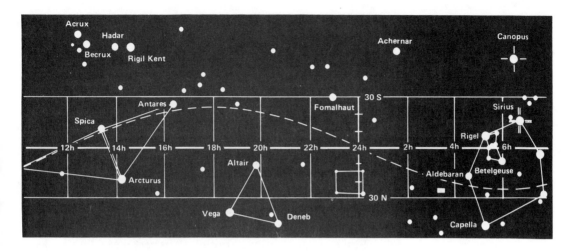

moment only one half of the celestial equator is above the horizon, the other half is invisible.

Which half? That depends only on the star time of the moment at your location. Suppose it is 12^h. Then the hour line labeled 12 on the equator of the strip map will be due North of you. If there were a star with RA 12^h, declination $0°$ (on the equator), it would be due North of you now, at the high spot of the equator you traced.

At the same star time of 12^h, a star on the equator with RA 18^h would just be rising at the East Point of your horizon. A star with RA 6^h would just be setting at the West Point of your horizon.

You can find the star time from standard time in Appendix A or calculate it from Figure 3.2. You can also estimate it from the stars you recognize in the southern sky. Turn the book until Figure 15.3 matches the sky; the number on top is your start time.

Now imagine the equator of the strip map forming an arch connecting the East and West points on your horizon and reaching its highest point north of you three spans and three fingers above the horizon, as calculated before. Mentally slide the strip until the present star time is highest, that's due north.

When you recognize some stars or star patterns, you can adjust the strip to match the sky without caring about star time as such.

Stars will never rise in the STZ if their north declination is greater than 90° minus your latitude.

For example, in latitude 23.5°S only stars in declination 66.5°N or less ever appear above your horizon. So the Big Dipper—Dubhe's declination is 62°N—will peek over your north horizon at times. The dimming near the horizon will dim its stars to third and fourth magnitude.

The first-magnitude stars at the north ends of our superconstellations—Capella, Vega, and Deneb—shine brightly here. They should stay recognizable as bright stars in southernmost Africa and central Australia.

Farther south, our superconstellations become badly crippled.

Sun, Moon, and Planets

The sun at noon in the South Temperate Zone is always north of you. For some readers in the northern hemisphere that seems crazy.

If your eyes, ears, and brains shut down at the first hint of mathematics—foreign currency, 24-hour time, metric measurements—you may like the following explanations. It contains no figures whatever.

- In the North Temperate Zone the celestial North Pole and Polaris are always and everywhere in the northern sky.
- In the NTZ the high point of the celestial equator, and the sun at noon, are on the opposite side, south.
- In the NTZ the celestial South Pole always and everywhere is below the South Point of your horizon—invisible.

If you have no problem with these statements, read on. Switch every mention of north to south, and the other way around. You'll get south to north.

- In the South Temperate Zone the celestial South Pole is always and everywhere in the southern sky.

- In the STZ the high point of the celestial equator, and the sun at noon, are on the opposite side, north.
- In the STZ the celestial North Pole always and everywhere is below the North Point of your horizon—invisible.

Let me reinforce that knowledge. In the South Temperate Zone the celestial South Pole is south of you. The high point of the celestial equator, and the sun at noon, are north of you. The celestial North Pole is north, below your horizon.

The best way to reinforce the sun at noon being north is a little story. It was written down, as far as we know, in the fifth century BC by Herodotus.

In the days of Pharaoh Necho (610–595 BC) a Phoenician fleet sailed from the Red Sea, around Africa, and returned by way of the Pillars of Hercules (today's Strait of Gibraltar) through the Mediterranean. That's about four times the distance of the route of Columbus to the New World.

The members of the expedition were ridiculed for the rest of their lives. Who but liars could make up a story of such a voyage? Who would believe people who claimed the sun was north of them when they sailed around southern Africa?

The scoffers were the ancestors of the crowds who later disbelieved Marco Polo, and their modern descendants who tell you the moon walks were filmed in a motel in Cocoa Beach, just south of the Kennedy Space Center.

To me it seems proof—more than 2,500 years later—that they had really been to southern Africa. Unless he had seen the sun north at noon, who'd come back with such an unlikely tale?

The celestial bodies that follow roughly the path of the sun— the planets and the moon—also generally are highest in the sky north of you. If they followed the sun exactly, there wouldn't be any exceptions. As it is, they stray at times a couple of fingers from the sun's path.

Near the Tropic of Capricorn that occasionally carries them a little south of your zenith. Without instruments we are not very good at finding the point exactly above our heads.

Even in the Bahamas the moon occasionally is on the "wrong" side of the zenith. None of the many sailors I talked to there had ever noticed it. Most thought I was telling a tall tale.

Mercury is a little less difficult to see in the STZ than farther north. Still it is only visible to the naked eye before sunrise until early dawn overwhelms it, or after sunset when it wins out over dusk. It is still low above your eastern horizon in the morning, and low over the western horizon in the evening.

The law of extinction near the horizon here too makes Mercury seem less bright than if it were higher up in the sky. But southern observers of Mercury are somewhat favored over northern ones.

In the STZ it is best placed for observation as a morning star when it reaches greatest elongation (west) from the sun in early April, as an evening star when it reaches greatest elongation (east) in early October.

Fixed Stars

The farther south you go the more northerly stars and constellations you'll lose.

Near the tropical border of the South Temperate Zone you'll lose every northerner's favorite asterism, the Big Dipper. About 10° latitude (700 statute miles, 1,000 km) farther south you may not find Deneb and Capella anymore. Technically they may still be above your horizon but far dimmer than their customary first magnitude.

There, say near Buenos Aires, Capetown, or Sydney, Hadar and Rigel Kent and the Southern Cross become circumpolar. They stay above your horizon all night, every night.

I picked these three cities, which happen to be in almost the same latitude (34½°S), not because they are good places to watch the stars. No city is. But in South America, Africa, and Australia you are likely to know if you are north or south of these places. Especially in Africa; if you are ½° south of Capetown you'd better start swimming.

Even before you reach the latitude of these cities, the two Magellanic Clouds become circumpolar. Both look like misplaced, far misplaced, bits of Milky Way. They go by different names.

Nubecula Major, literally the larger little cloud, is usually referred to as the Large Magellanic Cloud, or as LMC. Its center is near RA 5½^h, so it will be south and highest in your sky about an hour before Canopus, when star time is between 5^h and 6^h. The LMC will again be south, low above your horizon, when star time is between 17^h and 18^h. It is blob shaped, about three fingers wide and three high, and stays plainly visible on full-moon nights.

Nubecula Minor, literally the lesser little cloud, is usually referred to as the Small Magellanic Cloud, or as SMC. Its center is near RA 1^h, so it will be south of you and highest in your sky about one hour after Achernar, when star time is 1^h. The SMC will again be south, low above your horizon, when star time is 13^h. It is drop shaped, about two fingers wide and two high, and disappears from naked eye watchers on full-moon nights.

Each of these clouds, the nearest nebulous objects outside our own galaxy, is neither a nebula nor a cloud. Each is a galaxy: a large island universe that contains stars that emit light, and some that don't, globular clusters, nebulae, pulsars, black holes, and all the other bodies that make up any galaxy, including our own.

These two galaxies are thought of as members of a local group of about twenty such island universes. The best known member, besides our own Milky Way system, is M31, the spiral galaxy visible on dark nights in Andromeda.

Local seems a strange word to use for our neighbors. Light from LMC takes 160 thousand years to reach us, from SMC 190 thousand, and from the great Andromeda nebula two million years.

Perhaps these two island universes are so interesting because we have nothing like them in the northern hemisphere—nothing but the Milky Way, that disk with arms spiraling away from the center, made of two hundred thousand million stars, some brighter, some duller than our sun.

Seen from our slightly off-center place inside that system, the Milky Way is a spectacular band of diffuse light. Here and there the band breaks into two bands that braid around islands of darkness, or end abruptly.

Nowhere in its course around the night sky is there more variety in structures than in the STZ. Nowhere is that belt richer in jewels than in Scorpius, Crux, and Carina.

Here also is what looks like the darkest hole in the glowing band, the Coalsack, a three-by-four-finger-wide pear-shaped black dust cloud next to alpha in Crux.

Finding Directions in the STZ

Steering By a Star

Here, as everywhere else, it is more relaxing to steer a boat by a star ahead than it is to watch a compass all the time.

When the craft is on course by compass, look for a star dead ahead. You'll find it most relaxing when the star is a span or two above the horizon. When you have a choice of stars, pick one that's easy to recognize, one brighter than its neighbors or with another star nearby.

For the next little while steer for that star, then check your compass again. When the star has moved off, steer a little away from the direct line to the star. When it's too far off, try to find another guide star.

Don't let your knowledge of star motion mislead you. You may think you know that the stars seem to move from East through South to West. Here they move in much of the sky from East through North to West, but your guide star may run the other way.

Don't worry about it. Just believe your compass. Since you have little else to do at the time you'll probably wonder "How come?"

Think of a circumpolar constellation, say Crux. When it's high in the sky, above the pole, it moves as expected. When it's low in the sky, below the pole, it moves in what looks like the wrong way to get it back to the start for the next trip in the expected direction.

Gamma in Crux is on top when the Southern Cross is above the pole. Alpha Crucis is on top for the return trip. (In the northern hemisphere, Cassiopeia makes a similar turnabout, appearing as an M then twelve hours later as a W.)

You don't have to steer a boat to use this technique. You can use it walking, skiing, whatever.

Direction from the Stars

There is no star marking the celestial South Pole the way Polaris marks the North Pole. The star nearest the pole is fifth-magnitude Sigma in Octans.

But there's some help finding South from the Southern Cross. Acrux and Gacrux at the ends of the long arm of the Cross are in almost the same RA. When star time is about 12½^h or ½^h, the Cross will look straight up or straight upside down.

You can use that in reverse, without knowing local star time. Whenever the long arm of the Southern Cross is vertical, one end of it points to the celestial South Pole; the other end points due South. When equatorial stars rise or set they give you true East and West. See Figure 12.5.

Direction from the Sun

The best star, by far, for giving direction at rise or set is the nearest star, our sun. Without any equipment, without knowing the exact time, with only an approximate latitude, you can get an accurate direction from the sun. Just look up the direction where nearest date and your latitude cross in the tables in Chapter 12 (pages 126–129).

If you missed the exact moment of rise or set, or if your horizon is obstructed by hills or trees, don't worry. The figures given don't change more than 1° in the first ten minutes after sunrise or before sunset at the Tropic of Capricorn. From there the time for a 1° change gradually decreases to five minutes near the latitude of Cape Horn (47°S).

At noon in the STZ the sun indicates true North. Unfortunately only accidentally will noon be when an accurate clock in your area shows 12:00. To calculate the time of noon on a given day and place you have to apply some corrections.

• First you must correct for the sun being fast or slow compared to the mean sun. Only on four days a year does the sun actually reach its highest point at 12:00 noon. The rest of the time it is up to one-quarter hour fast or slow. The Local Time of Noon table on page 130 gives the time of noon at the standard meridian of your time zone. For example, 60°W for much of South America.

- To calculate the time of noon you must know your longitude accurately. For every degree of longitude you are west of the standard meridian, noon for you will be four minutes later. For every degree of longitude you are east of the standard meridian, noon for you will be four minutes earlier.

  If that is not good enough for your work, you must get your longitude to the nearest ¼° (15 minutes of arc) to get the error down to one minute in time.

- Finally you must add one hour—usually—when daylight saving time is in effect. Say your first two steps placed noon for you at 12:08. The sun won't be north of you until your watch, set ahead for daylight saving (sometimes called summer) time, shows 1:08.

If that sounds too complicated, you also can get North or South by shadows of equal length—one before, one after noon.

At some time before noon mark and measure the end of the shadow cast by a vertical pointed pole on a flat, level surface. A string hitched around the base of the pole remembers the length you mark with a nail or thread or tape.

After noon watch the shadow. When it becomes almost as long as the morning shadow, test it at short intervals. Mark the point where the shadow falls when it's as long as the morning shadow. North lies in the direction of the center of the pole from the mark halfway between the morning and afternoon marks. Use string also for finding the half distance. Just fold the string—now marked for the whole distance between the shadow marks—in two.

Warning

You may have heard or read of two methods that sound logical and simple.

One uses the hour hand of your watch and the sun to find North. In the STZ that method doesn't work. If you want to know why not, look back to "South from Your Watch" in Chapter 12.

The other method supposedly lets you find East from the westward movement of a shadow. Here, as in most of the world, this method during most of the day is badly misleading.

Direction from a Moving Shadow

This method has you watch a shadow cast by the sun on a flat, horizontal surface. As the sun moves west, the shadow of the tip of a vertical pole is supposed to move east.

That sounds like a simple way of getting true East, from which you can get any other direction. Unfortunately it's not so.

In the morning the sun rises higher in the sky, so the shadow as it moves west in the STZ also moves north. In the afternoon, the sun sinks back toward the horizon, so that shadow also moves south. At times in the STZ the north or south movement of the shadow is greater than the easterly trend.

Near noon the sun rises or sinks very little, but moves west at a rapid rate. That is the time the method is useful. Within forty minutes, either side of local apparent noon—roughly between 11:20 and 12:40 sundial time—mark such a shadow, then perhaps half an hour later mark it again.

Your guess of correct noon can be quite a bit off, and you'll still get into the critical period. The shadow can be cast by a monument onto a public square, or a ski pole pointed end up.

The second shadow mark will be east of the first one. The accuracy will be best in local winter, March through September. The calculated error near the Tropic of Capricorn will be less than 4°, near the Antarctic Circle half that.

Finding the time of local noon from the shortest shadow at that moment is not practical. The length of the shadow near noon barely changes, while its direction moves rapidly. That is, after all, what makes it the best time for finding shadows moving mostly east.

16

The Antarctic

Technically the Antarctic Zone starts at the Antarctic Circle, near latitude 66½°S. It ends at the South Pole, which is by definition in latitude 90°S. From any point on the Antarctic Circle to the South Pole is a distance of more than 1,600 statute miles (2,600 km).

The Antarctic Circle encloses virtually the whole of the continent of Antarctica. The largest part left out is the north end of the Antarctic Peninsula and a few nearby small islands. None of the nearest continents—South America, Africa, and Australia—comes within 700 miles (1,100 km) of the Antarctic. South America's Cape Horn is nearest.

You could define the Antarctic as the land of the southern midnight sun. That reminds you that it is the southernmost part of Earth and that the sun at midnight is south. As everywhere in the southern hemisphere the sun at noon is north of you. So when it doesn't go below the horizon at midnight—twelve hours later, and halfway around the world—it is south of you.

At the Antarctic Circle on December 7 the upper edge—limb is the astronomer's term—of the setting sun just touches the horizon when it begins to rise again. The sun will not set again until the evening of January 5. These dates may shift by a day due mainly to the same effect that makes leap years necessary.

It may seem strange that for about four weeks the sun stays up all night only as far as the Antarctic Circle.

Near the winter solstice—the time you are now reading about—the sun's declination changes very little from day to day.

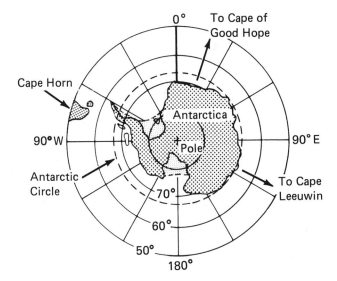

16.1   The Antarctic.

Within the dates given it changes only about ¾ of 1°. (The sun's diameter is about ½°.) At the equinoxes it changes that much in 48 hours.

Fixed Stars in the Antarctic

All the practical astronomy from the South Temperate Zone, Chapter 15, works near the border of the Antarctic. But if you are transported suddenly to the southern hemisphere, you'll have to unlearn a few basic facts. South of the equator the sun at noon, the stars, planets, and moon when they are highest in the sky are north of you, not south as north of the equator.

Recall that wherever you are our reference line—the celestial equator—goes through the points due East and West of you. Its highest point in south latitudes is due North of you.

The celestial equator is 90° minus your latitude degrees above your north horizon. On the equator (latitude 0°) that is 90° above your horizon, directly overhead. On the South Pole (latitude 90°S) it is 0° right on your horizon.

You can demonstrate what happens when you go south of latitude 0°. Hold this book as if you were reading it. Now raise

185

it to simulate what happens to the celestial equator when you travel south.

When you can't raise it any higher the book, now overhead, has arrived at latitude 0°. To see what happens when you keep going, you have to cheat a little because you can't see behind you. Have someone else hold the book while you turn around. Or hold the book a little away from you with only one hand.

Either way, you'll see the print upside down. But the eastern margin is still east, the western margin west. Or instead of the book use a map with south on the bottom and east on the left as usual. What will you have after you pass it overhead? An upside-down map with North on the bottom and East still to the east, West to the west.

What that means for naked-eye astronomy in the southern hemisphere is this:

- The high point of the celestial equator is above the North Point of your horizon.
- The sun, moon, planets, and stars, all of which move in one day virtually parallel to the celestial equator, reach their highest point north of you.
- They seem to rise in the east and set in the west just as they do north of the equator.
- Constellations may seem upside down to northerners.
- Perhaps most surprising, a waxing moon phase looks like a waning one, a first quarter like a last quarter, and the other way around.

To find the high point of the celestial equator in the southern hemisphere you subtract your latitude from 90°. (Just as in the northern hemisphere.)

On the Antarctic Circle (latitude 66½°S) the high point of the celestial equator is then 23½° above your horizon. Face due North and measure one span and two fingers from the sea horizon.

If distant mountains obstruct your horizon, guess where the sea horizon would be. On the plateau you won't be far off by measuring from where the sky touches the flat land.

You may have to use a compass to find true North. Find out

what the local correction is and apply it. In some parts of the Antarctic magnetic compasses are unreliable, or worse. That's due to the proximity of the magnetic South Pole, which lies near the coast of Wilkes Land.

The celestial South Pole, which would let you find true North, is not marked by a bright star, as the North Pole is by Polaris. Often a good way to find North is from local star time and a star or constellation you happen to recognize.

Appendix A lets you find star time for any time of any day of any year. For example, for June 7 at midnight star time is 17^h. Star time here equals RA^h of the meridian now due North of you.

Is there a star near the 17^h line on the star map (Figure 15.4)? Indeed there is: Antares at 16.5^h, the brightest star in Scorpius. Look for it in the sky. A scorpion, which this constellation really resembles, looks about the same with either end up.

Face Antares now. The high point of the celestial equator on the Antarctic Circle came out to be 23½° above your horizon. The table of bright stars (Appendix E) gives the declination of Antares as 26°S. So Antares will be 26° above the equator.

Now find the spot 26° (one span and three fingers) below Antares. Give that spot a Roman salute, then swing one arm left, back to the spot saluted, and on to the right. You have just traced—west to east—the celestial equator in your sky.

Now align the equator on the strip map for the southern hemisphere (Figure 15.4) with the line traced in the sky. Since star time now is 17^h, the space between the hour line labeled 16 and the one labeled 18 should be at the high point and close to true North.

That works the other way around, too. If you have true North from a compass (corrected for local error), star time, and approximate latitude, you could trace the equator in the sky and identify the stars from the star map.

Often the stars in or near the southern cap (Figure 16.3) help identify the stars in the Antarctic. Imagine the center of the cap, the celestial South Pole, above the point south of you on the horizon. The center will be as many degrees above the horizon as you are south of the equator. (On the Antarctic Circle that'll be 66½°, or about three hands and three fingers.)

Turn the map until you get a match with the brightest stars you see. The numeral now on top is your local star time.

That will help you fit the strip map. The vertical line labeled with star time will be north of you, with the equator 90° minus your latitude degrees above the horizon. (On the Antarctic Circle 90° − 66½° = 23½° or one span and two fingers.)

When you travel farther south the high point of the celestial equator—north of you—will drop 1° as your latitude increases 1°.

The celestial South Pole—here south of you—rises 1° higher above your horizon for every degree your latitude increases.

Your latitude will increase (decrease) 1°
when you travel south (north) 60 nautical miles,
which is about 69 statute miles (111 km).

As your latitude changes and the South Pole rises, more and more stars become circumpolar, i.e., they never set.

Already at the Antarctic Circle, Antares technically stays above your horizon all the time. But when it is within an hour of being south of you it will not look its usual bright self due to an attack of extinction on the horizon.

But you'll never see the first-magnitude stars Capella, Pollux, Vega, and Deneb.

Sun, Planets, and Moon

In the Antarctic the sun never completely sets for about four weeks, between December 7 and January 5.

The moon and the planets closely follow the annual path of the sun, so they too will occasionally stay above your south horizon all night.

At the South Pole

Let me take you quickly to the extreme of the Antarctic Zone, the South Pole. I suspect that few readers will go there just to watch the heavenly bodies. But knowing what things are like

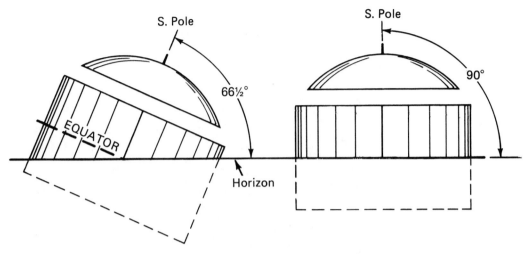

16.2   Dome of the sky in the Antarctic: the south cap and the strip map. Left: in place at the Antarctic Circle, the northern border of the Antarctic. Right: in place at the South Pole.

at the South Pole will give you a good idea what must happen between the northern border (the Antarctic Circle) and the pole.

At the pole the celestial equator falls on the horizon.

Therefore all stars in the northern—that is lower—part of the strip map (Figure 15.4) are below the horizon, invisible at all times.

The stars in the southern—that is upper—part of that map, or attached to it, will always be above your horizon at the pole.

They will always circle the sky at the same distance from the horizon. The distance above the horizon of each star will equal its declination.

The center of the southern cap is directly above you. You can think of the entire cap as suspended from your zenith. You can align the stars on the cap with the ones you see in the sky. Then align the star times with the ones of the strip map.

The sun's path is in the northern part of the map about half the year. So the sun then is permanently below the horizon. It is night, or twilight, around the clock for about half the year. To make up for that the sun doesn't set from about September 22 to March 22. The midnight sun lasts an average of 183 days.

16.3   The southernmost stars.

Some readers comparing these dates with the ones at the North Pole may question my arithmetic, or suspect a misprint. Not guilty: The southern hemisphere is being shortchanged by about a week's sunlight.

But the sun makes up for it. When it shines at the South Pole, it is nearer Earth than when it shines on the North Pole. The sun is nearest in the first week of January, most distant in the first week of July. (Again no author's goof, no misprint.)

The moon spends about half of each month in the northern part of the strip map and out of sight.

The planets, which travel nearly the same path as the sun, are also out of sight when they are in north declination. You can check the declinations of Mars in the calendar in Appendix B.

Jupiter stays invisible here during about half its trip around the sun, which takes almost twelve years. You won't see it here from March 1987 to October 1992, and from March 1999 through late 2004.

Saturn stays invisible for about half its 29-year trip around the sun. You won't see it here from mid-February 1997 through 2000.

Elsewhere in the Antarctic Zone these planets, and Mars, will rise and set when the numbers of their declination (N) plus your latitude (S) add up to less than 90°. That's the theoretical limit. To allow for extinction on the horizon get the declination from Appendix B and add your latitude. You should have no trouble finding the planet rise and set if the figures add up to less than 80.

For example, can you see Mars or Regulus in declination 12°N from latitude 80°S? No. From the Antarctic Circle (66½°S)? No problem. From latitude 76°S? Marginally, and dimly on the south horizon, near star time 10^h.

Finding Directions in the Antarctic

The Antarctic is very different from all other zones when it comes to finding directions from the sky. Several of the methods used elsewhere don't work here. But you can use two others that don't work in the temperate zones and the Tropics. One method works only here.

Steering by a Star

Here, as everywhere else in the world at night, you can steer a boat, walk, or ski a given course easier than by watching a compass all the time.

When you or your craft are on the desired compass course, look for a star directly ahead. A star not too close to the horizon but comfortably low in the sky is best. At least one span and no more than two spans above the horizon seems most restful to most people.

You don't need to know the name of the star, or its constellation. But it's helpful if there is something special about the star so you can find it again after taking your eyes off it. A nearby star may help. It may be left or right of it, above or below, kitty-corner, much brighter or dimmer.

Steer for your star for a while. Then check the compass again.

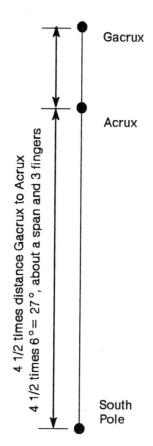

16.4 *Finding the celestial South Pole from Crux.*

If you are still on course, continue. At the next check your guide star may have drifted left or right. If so, steer a bit right or left of it.

How fast do stars or planets drift in this area? Here's a rule of thumb: A star or planet one or two spans above the horizon will drift less than one finger (2°) in eight minutes.

On one of your next course checks, find a new star and repeat the procedure.

Directions from the Stars

As everywhere else, stars on the celestial equator here rise due East and set due West.

Whenever you see the stars almost on the equator in the constellations Orion, Virgo, and Aquila rise or set, you'll have a good East or West.

Figure 12.5 shows how to recognize these stars. They are easier to recognize when they are setting, after you have watched the entire constellation earlier. For rising, star times help.

As you approach the South Pole these stars don't rise sharply at a given moment but seem to drag along the horizon. The much diminished brightness near the horizon doesn't help either, but the method is useful in the northern Antarctic.

Sorry, the southern hemisphere lacks a bright star near the celestial South Pole. That would give us an easy way to find it.

At times the Southern Cross helps. The stars that form the long arm of the cross, Acrux and Gacrux, almost point to the celestial South Pole. Whenever the cross is upright (star time 12½^h) its long arm lies on a line that runs almost North–South. Acrux, the brighter of the two stars, is the one nearest the South Pole.

By the way, if you ever want to find Sigma in Octans, the star closest to the pole—an area almost devoid of stars—here is how (in the dark of the moon). The distance from Acrux to Gacrux is 6° (three fingers). The distance from Acrux to the pole is 4½ times that much, or 27° (one span and three fingers). When Crux is upright, above the pole, measure that distance in the direction away from Gacrux.

The Gacrux through Acrux line so measured misses the pole

by a little more than two fingers. A line that long, at a right angle to the main line on the Alpha and Beta Centauri side, will hit the pole. Sigma Octantis (magnitude 5.5) is a half finger farther on the same short line.

Directions from the Sun

Sunrise and sunset are poor guides here. To get useful bearings, the tables for high latitudes would need far more lines for dates and columns for latitudes than elsewhere.

The sun at noon is useable all the way to the pole, but it needs accurate time and two corrections, sometimes three: for the time of noon on a given date; for the difference in longitude between your location and the standard meridian; and for daylight saving time when in use (see the local time of noon table in Chapter 12).

The sun before and after noon method uses the equal lengths of shadows. It works here whenever the sun at noon is high enough above the horizon. You don't need a watch or knowledge of your longitude. But you must guess correctly that it's still before noon.

Before noon you measure the length of the shadow a pointed vertical object casts on a level, flat surface. Mark the point. When the afternoon shadow reaches the same length, mark that point.

Measure the length of the shadow with a string looped loosely around the base of the pole that casts the shadow. Use the same string later to measure the distance between the forenoon and afternoon marks.

North lies on a line from the center of the pole to the point halfway—fold the string—between the marks.

The hour hand of a watch and the sun whenever it shines work nicely in high latitudes in the Arctic. Here the sun still moves at half the rate of the hour hand. But in the Antarctic it seems to turn in the opposite direction of clock hands. That spoils the simplicity of this method of direction finding.

17

Celestial Navigation

This chapter is not about naked-eye astronomy. Consider it a bonus for having read this far. It sketches a method of navigation (by objects in the sky) that at first glance looks like magic.

You squint through a sextant at, say the sun, twirl a knob, and note the time. Then you look up a couple of figures in an almanac. You then do a little adding and subtracting, and look in another book of tables. A little more adding and subtracting. Then you draw a line on a map of your area.

You announce, "That's where we are." And usually you'd be correct to within a mile or two.

The same method for all the celestial bodies: sun, moon, planets, and fixed stars. And it works anywhere in the world.

How does it work? Think of Earth as a sphere inside a much larger sphere to which cling all the celestial bodies. Any of these bodies, say the sun, at any given moment will be directly above one point on Earth. At that point—let's call it the Ground Point, or GP—you would be standing on your own shadow.

At the GP the angle between the center of the sun's disk and the horizon—the line where ocean and sky meet—is exactly 90°. The same rule fits the moon. For stars, forget about the disk. The angle between star and sea horizon, a star's altitude, is 90° at the GP.

Now move away from the GP, in any direction, until the altitude becomes 89°59'. There if you measured your distance from the GP, you'd find it to be one nautical mile. When the altitude is 89°55', you'd be five nautical miles from the GP.

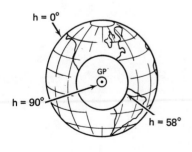

17.1 The world seen from directly above the Ground Point. There the sun's altitude is 90°. Along the outer circle the altitude is 0°. The sun is on the horizon, as at sunrise and sunset. The inner circle connects all points where the sun's altitude is 58°.

h = 0°

GP

h = 90°

h = 58°

Where it is 89°00', you'd be 60 nautical miles from the GP, and so on.

That simple relation (one minute of arc per nautical mile) makes navigators of all nations use the same measure at sea, for what on land they'd call about 6,076 feet, or exactly 1,852 meters.

The *Nautical Almanac* (available at any U.S. Government bookstore; call 202-783-3238) lets you find for every second of the current year the location of the GP of the sun, moon, Venus, Mars, Jupiter, Saturn, and the fifty-seven navigational stars.

A marine sextant lets you measure the altitude of any of these bodies. A chronometer lets you get the exact time of that observation.

Let's look at an example with nice round numbers. Your observed altitude is 58°00' at a time when the sun, according to the current *Nautical Almanac,* is directly above 15°00'N, 90°00'W. Your observed altitude is 32°00' less than at the Ground Point. You are 32 times 60 or 1,920 nautical miles from the GP.

With what we know so far about this sight, you could be anywhere on a circle with a radius of 1,920 nautical miles and centered on the GP. You could call that circle a circle of equal altitudes. It connects all points where at this time the sun is at an observed altitude of 58°00'.

A navigator might call it a *position line.* By that he means that he is on that line, but does not know just where on the line. Here's a landlubber's equivalent: "On the Florida Turnpike" is a position line. You'd be somewhere between Miami and Wildwood, more than 200 miles north. Add "and State Road 60" and you have a fix. You have pinpointed your position: where State Road 60 crosses the Turnpike—at Yeehaw Junction.

There are no marine charts (seagoing maps) that let you draw such a large circle without distortion.

You might think you could plot it on a globe. But a suitable globe would have to be twenty feet (seven meters) in diameter. Not practical.

But you are not interested in the whole circle. Because you left Boston in your yacht before breakfast, you are certain that

17.2 The circle of equal altitudes enlarged. Four navigators—near Boston (Massachusetts), Georgetown (Guyana), Lima (Peru), and Los Angeles (California)—observed the sun when it was directly above the Ground Point (15°N, 90°W). The arrows pointing toward the Ground Point show the bearings of the sun at the four assumed positions.

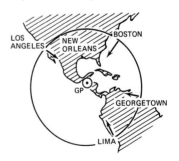

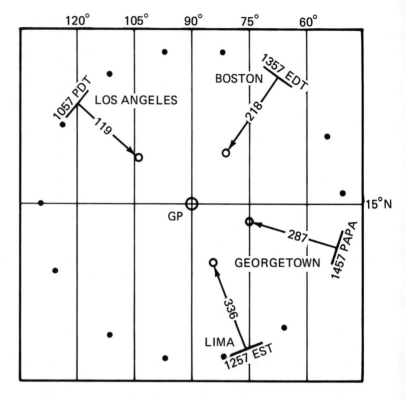

17.3 Position lines drawn, at right angles to the bearings, at the four assumed positions. The labels are the civil times at these places. Navigators will use Greenwich Mean Time, 1757, at all four locations.

you are not at lunch time near Georgetown or Lima or Los Angeles.

So you use the first magic trick: You assume a position near where you are likely to be. You base that position on your last known position and the movement of your vessel since then: in our example, you may assume 42°N, 70°W. The choice is not critical. You get the same result if you use 41°N or 43°N and 69°W or 71°W. That means that here you can assume a position anywhere in an area of about 11,000 nautical square miles.

With the assumed position and the data found earlier from the *Nautical Almanac* for the sun on that day at that time, you enter the second book, the *Sight Reduction Tables* (also available from any U.S. Government bookstore).

In our example you find the sun's bearing to be 218°, roughly southwest. That's reasonable for the sun about two hours after noon. Since the sun by definition is above the GP, that is also the bearing of the Ground Point from the assumed position. That lets you plot the bearing of the GP from the assumed position.

Now you use the second magic trick: You draw a small part of the circle of equal altitude as a straight line at right angles to the bearing. However far you are from the GP, the center of the circle, you can draw that short straight line on your local chart.

Suppose for now that the altitude printed in the sight reduction tables happens to be exactly the same as your observed altitude. Draw a line through the assumed position at right angles to the bearing. That's your position line at the time of observation.

What you did is this: By plotting the bearing, you drew the radius of the circle of equal altitude that ends at your assumed position.

Geometry teaches that the periphery of a circle at any point is at a right angle to the radius that ends at that point. The short line you drew at right angles to the bearing is a nearly correct rendering of the circle of position (or equal altitude) near the assumed position.

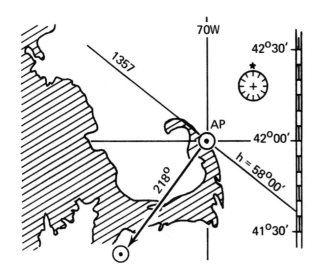

17.4 Position line drawn exactly through the assumed position because the observed altitude of the sun happens to be the same as the altitude calculated in the Sight Reduction Tables.

How nearly correct? In our case the radius of the circle is 1,920 nautical miles. Thirty nautical miles on either side of the assumed position our straight line is ¼ nautical mile away from that circle.

In plotting our position line we assumed the observed and calculated altitudes to be the same. That will happen only by accident.

What if the observed altitude is smaller than the one given in the tables? That means you are farther away from the Ground Point than you had assumed. For every minute difference, you are one more nautical mile farther away from the GP.

Say the observed altitude is five minutes smaller than the one calculated for the assumed position. You'd draw your position line through a point five nautical miles farther from the GP than the assumed position.

Now suppose the observed altitude is five minutes greater than the one calculated for the assumed position. You'd draw your position line through a point five nautical miles nearer the GP than the assumed position.

How useful is a single position line? In the illustrated example a single position line tells you what to do.

Suppose your observed altitude turns out to be 58°00′. If your strategy was to give Cape Cod a wide berth you'd better

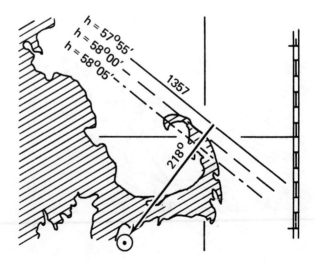

17.5   *The position line drawn dashed repeats Figure 17.4. The solid line is for an observed altitude five minutes smaller than what the* Sight Reduction Tables *give for your assumed position. The dot/dash line is for an observed altitude five minutes greater than in the tables.*

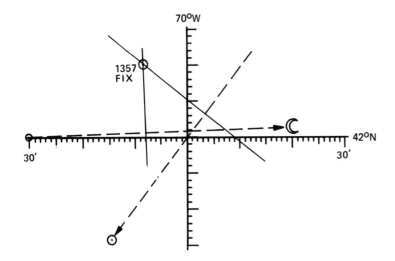

17.6 Two-body sight. A moon sight (worked from a slightly different assumed position) is combined with the sun sight in Figure 17.5. For clarity the bearings of sun and moon are shown as dashed lines, position lines as solid lines. The crossing point of the two position lines marks your fix.

steer a little more northerly course. If you were aiming for Provincetown, inside the hook of Cape Cod, you'd better get in a little southing.

On many days each month the moon is visible for several hours in daylight. At those times you can take a sight of the moon right after shooting the sun.

You plot the two position lines on your chart. Since your vessel is on both these lines at the same time, it must be where the two lines cross. You have a fix, a definite position.

You can also use a third magic trick: You can take two sights of the sun, say three hours apart. When you have plotted the second position line, bring the first one forward to allow for the travel of your vessel between sights.

The plot will look as if the vessel had towed the line on the course and at the speed made good between the sights. Again you have a fix where the two position lines cross.

I can't blame you if you think that's on the level of Baron Munchhausen pulling himself and his horse out of a swamp by his pigtail. But for more than two centuries combining two such sights of the sun has been the standard method for navigating the ships of all nations.

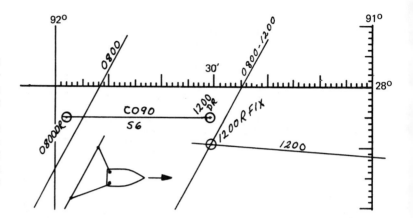

17.7 Running fix. From the 0800 dead reckoning (DR) position, near the 0800 position line, a vessel sails on a course of 090°, at a speed of six knots (six nautical miles per hour).

A sun sight is worked at 1200. The navigator now moves the 0800 position line forward as if the vessel had towed it. Where the "towed" position line crosses the noon position line is the 1200 running fix.

That and one other sight: star sights. To take such a sight with a marine sextant, you need to see the stars and the sea horizon. So the time for star sights is limited to short periods at dawn and dusk.

Two star sights taken at virtually the same time give you a fix. But most navigators prefer to shoot three stars rapidly one after the other. That gives three position lines that should form a very small triangle—known as the cocked hat—around your position at the time of observation.

Volume I of the *Sight Reduction Tables* makes it simple to take and work three star sights, almost at once.

First you estimate, or calculate from data in the *Nautical Almanac,* the time for your star sights. You can also wait until it's almost light enough to see the horizon in the morning. Or dark enough to see the first stars in the evening.

With that approximate time and your longitude, get the Local Hour Angle (LHA) of Aries from the *Nautical Almanac.* That five-word term is just another measure of local star time, which you've used all through this book.

Next open Volume I to your latitude—anywhere between 89°N and 89°S. On the line corresponding to the LHA of Aries, you'll find data for seven bright stars available for observation. Star names in capital letters indicate first-magnitude stars. Three of the seven stars are marked by asterisks to indicate they are at this time well placed for a three-star fix.

You don't have to rely on your knowledge of the stars. You can find them from the data. For each star, the table gives its calculated altitude and true bearing at this time.

For example in latitude 28°N when the LHA of Aries is 143°, the book suggests for a three-star fix:

*ARCTURUS   *CAPELLA   *SIRIUS

For Arcturus it gives calculated altitude 26°27', true bearing 81°. To find Arcturus set your sextant to about 26½° and face East. (East is at 90°, so 81° is a little left of that.) Sweep the horizon. The first bright star you catch in the telescope will be Arcturus. Adjust the sextant until the star kisses the horizon.

Write down the exact time, then write down the sextant reading. Go on to Capella and finally to Sirius. A few minutes later—after adding and subtracting small corrections for each of the three stars— you can plot your position on the chart.

The method I sketched for converting sextant measurements into position lines and fixes was slow in coming.

Ship's officers struggled with spherical trigonometry and logarithms for more than a century. Then, in 1837, Captain Thomas H. Sumner, a Harvard graduate, discovered the position line.

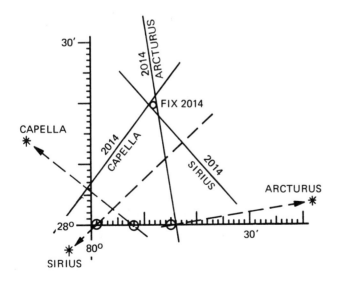

17.8 Three-star sight. Bearing lines are shown dashed, position lines solid. The three position lines enclose the fix.

About forty more years passed before Commander Marcq Saint–Hilaire of the French Navy introduced the altitude difference method. That's the method that has you move your position line one mile for every minute difference between computed and observed altitudes.

The Basics: A Summary

If you have stayed with me this far, you know how celestial navigation works. Here I summarize the basics, then I'll fill in some details.

- You measure the altitude of a celestial body. That is the angle between sea horizon and the sun, moon, planet, or star.

- You note the exact time of the observation.

- In the *Nautical Almanac* you find the point on Earth where the celestial body was directly overhead at that time. There— at the Ground Point—the body's altitude was exactly 90°.

- The closer your observed altitude is to 90° the nearer you must be to the GP. The smaller your observed altitude, the more distant you are from the GP.

  Observers who get the same altitude when measuring the same body at a given moment must therefore be at the same distance from the Ground Point. They must be positioned somewhere on a circle centered on the Ground Point, a circle of equal altitude.

- To narrow down your place on that circle of equal altitude, first choose a position near where you are likely to be. That normally would be your last position moved forward at the course and speed of your vessel.

- With that assumed position and the data from the *Nautical Almanac,* you enter the *Sight Reduction Tables.* There you find the bearing of the Ground Point from your assumed position. You can draw that directly on the chart of your area, just as you'd plot, say, the bearing of a lighthouse from your vessel.

- The tables, next to the bearing, also give the altitude of the body at the assumed position. Compare that with your observed altitude.

  If they happen to be the same, draw a line through the assumed position at right angles to the bearing line.

  If the observed altitude is smaller, you are more distant from the Ground Point. Plot the line not through the assumed position but through a point farther from the Ground Point, one nautical mile farther for each minute of altitude difference.

  If the observed altitude is greater, you are nearer the Ground Point. Plot the line not through the assumed position but closer to the Ground Point, one nautical mile nearer for each minute difference.

- The short line you just drew—at right angles to the bearing— approximates closely the circle of equal altitude. At the time of the sight your vessel was somewhere on that line, or very near it. But you don't know just where on the line.

  Such a single position line is often useful. It may identify a distant landmark, check on your course, or your vessel's progress.

- More often you'll want a precise position, a fix. You can get that by taking—within minutes—sights of two bodies, say sun and moon. Your fix is where their position lines cross.

- When the moon isn't available, you combine the position lines of two sights of the sun taken several hours apart. By allowing for the travel of the vessel between sights you get a running fix.

- At dawn and dusk stars give more accurate fixes. Most navigators shoot three stars in rapid succession. The third star serves as a check.

Some Fine Points

The sextant, invented in 1730 independently in England and America, accurately measures the angle between the sea horizon and a celestial body.

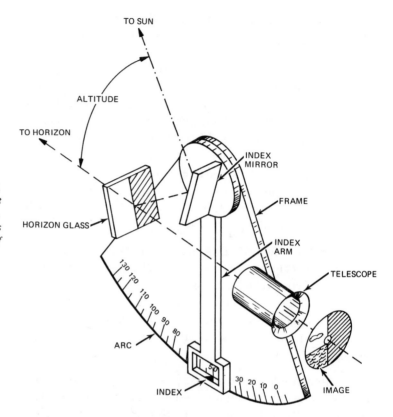

TO SUN

ALTITUDE

TO HORIZON

INDEX MIRROR

FRAME

HORIZON GLASS

INDEX ARM

TELESCOPE

130 120 110 100 90 80

ARC

50

30 20 10 0

INDEX

IMAGE

17.9 Marine sextant (simplified). The shade glass needed to protect your eye from the sun is not shown. A micrometer on modern sextants lets you read the altitude to a fraction of a minute.

On land, levels, plumb lines, and bowls of mercury had been used in measuring the altitude of stars since antiquity. On a ship, which rolls and pitches, none of these works.

The cross-staff, the first instrument to use the sea horizon, made the navigator look at the sun and the horizon at the same time, a nearly impossible task.

In the sextant it's all done with mirrors. Two mirrors, as Isaac Newton suggested in an unpublished letter thirty years before the first sextants were built.

Looking through a telescope, you see the horizon through the clear (left) part of the horizon glass. Face the sun. Snap the shade glass in place to protect your eye. Then pivot the index arm until you can see the sun—reflected by the index mirror— in the silvered (right) part of the horizon mirror. Then gently move the index arm, with its attached mirror, until the sun seems to touch the horizon.

Note the exact time. Then read the position of the index on the arc. That is the sextant altitude of sun at the time of your observation.

The sun is by far the body most used in navigation. Weather permitting, it's available all day long. Also you can catch the sun in a break of the clouds, and even through a thin cloud cover. Stars have to be shot during a short part of dusk or dawn.

A moon sight in daytime can be combined with a sun sight. For such a fix the moon's bearing must make a good angle with the bearing of the sun. At night moon sights are unreliable. Moonbeams reflected on the water often create a false horizon.

Stars, away from moonbeams, sometimes permit a sight at night. During dawn and dusk, when the horizon is visible, stars give you great fixes.

Almanacs list fifty-seven navigational stars. They include all the first-magnitude stars except Becrux. (Its neighbors Acrux and Gacrux are listed. Polaris is not included among the navigational stars. You can work it without sight reduction tables, using data that take only three pages in the *Nautical Almanac.*) The remaining thirty-seven stars include a handful of third-magnitude stars in desolate areas of the sky.

You can work star sights in all oceans, in every season, using only about twenty stars.

The planets Venus, Mars, Jupiter, and Saturn, listed in the daily pages of the *Nautical Almanac* (Mercury is not), are favorites of some navigators. Brighter than fixed stars, they can often be shot when the horizon is still well lit in the evening, or when the stars have already faded in the morning.

Accurate time is essential in celestial navigation. Timepieces don't have to be as accurate anymore since we can get radio signals worldwide, twenty-four hours a day. Quartz watches checked daily with the radio can serve in place of the three chronometers required at the beginning of this century. (If two chronometers disagreed, you needed a third to tell which was correct.)

Almanacs use Greenwich Mean Time, also called Universal Time or, in the Navy, Zulu. The radio time signals use the same time. It's smart to keep your navigation watch on the same time and date to avoid having to subtract or add for zone time, possibly some local time, and perhaps daylight saving time. Almanacs use twenty-four hour notation, so navigators also work in what you may know as military time.

Small corrections are needed. That's the little adding and subtracting mentioned in the second paragraph of this chapter. The sextant altitude, for example, needs these corrections:

- For dip of the horizon, the difference between the visible horizon and the horizontal plane at your location. The earth's curvature lets you see beyond the true horizon. The higher your eye at the observation, the more the visible horizon dips. (That's why you see farther from a tower than from the ground.)

  In the cockpit of a sailboat you may have to deduct 2' from the sextant altitude; on the bridge of a cruise ship 8'.

- For refraction, the bending of light rays passing through the atmosphere. A star ten degrees above the horizon will appear 5' higher than it would be without the atmosphere.

  The higher the star, the less its displacement. The refraction becomes zero for a star directly overhead.

- For the semidiameter of the sun's disk. The position of the sun is calculated for the center of the disk. In a sextant we bring the low point—called lower limb—in contact with the horizon.

  The correction for the difference between the altitude of the observed point and the center of the disk is about 16' to be added to the sextant altitude.

- For the semidiameter of the moon's disk. Allow as for the sun. When the lower limb of the moon is not touched by sunlight, correct for the upper limb.

- For the parallax in observations of the moon. The position of the moon is calculated for an observer at the center of Earth. You measure the moon's altitude from some point on the surface.

The parallax is the difference in angle between two such observations. It is greatest—57' on average—when the moon is on the horizon.

You'll find all these corrections on the inside covers and fly-leaves of the *Nautical Almanac*. The ones most used are also printed on a card you'll find handy as a bookmark.

Using them is easier than this list makes it sound.

Take the sun sight as an example. You probably take all your sights from the same level. That makes the correction for height of eye—the dip—always the same. Without really trying you'll soon remember that correction.

That leaves only the lower-limb correction, which here is combined with the one for refraction. You will almost always need the correction for the lower limb and altitudes greater than 10°; you'll find it in the first column of the inside front cover and on the bookmark.

Sights of Jupiter, Saturn, and the fixed stars are just as easy to correct; the same dip as for the sun, and a correction for refraction.

Venus has her own correction of a few tenths of a minute of arc. For months on end you can safely forget about it; the same for Mars. You'll find all these corrections in the center column of the inside front cover and on the bookmark.

Earlier you read that the *Nautical Almanac* lets you find data for every second of the current year. Perhaps that made you think it had entries for every second. You can be glad it doesn't. In the same format as the present almanac, it would fill more than 900 volumes every year.

After the proper corrections are applied, sextant altitude becomes observed altitude. That's the altitude you compare with the one calculated for the assumed position in the sight reduction tables. The difference in these altitudes tells how far—toward or away—you are from the assumed position. (See Figure 17.5.)

The present almanac gives data for every hour of the current year. It also shows, at the bottom of the columns, the hourly changes. Near the end, printed on yellow paper, in sixty columns of sixty lines it gives the correction for every minute and second for any hourly change used in the almanac.

After a few times, finding the values in the yellow pages and adding them to the hourly data takes less than a minute.

Volumes II and III of the *Sight Reduction Tables,* H.O. 249, for sun, moon, and planets, use a similar approach. They give calculated altitudes for whole degrees of declination. One-page tables at the end of each book (and on bookmarks) give corrections for the minutes of declination.

Volume I, used for stars, shows altitudes and bearings calculated for whole degrees of latitude and local hour angle. You assume a position near where you probably are, and where latitude and local hour angle are degrees without minutes.

You see, it isn't really magic!

Appendices

A: Star Time from Local Time and Date
B: Calendar for the Moon and Four Planets
C: Periods of Visibility for the Planet Mercury with Magnitude
D: Modern Climate Zones
E: Brightest Stars and Their Constellations
F: Key to Pronunciation

Key for Appendices B, C, and E

-o-   too close to sun for observation

-o- to 5   invisible from beginning of month to about 5th

-o- after 25   invisible from about 25th to the end of month

AM   visible on the morning, before sunrise, in the eastern sky

PM   visible in the evening, after sunset, in the western sky

RA^h   right ascension, in hours and tenths. The left/right position of a celestial body (see Figure 8.2); also the star time when the planet or star is highest in the sky, due south in the northern hemisphere. RA^h for Jupiter and Saturn is for middle of month.

DEC°   declination, in degrees north (N) or south (S) of the celestial equator. The up/down position of a planet or star (see Figure 8.5). DEC° for Jupiter and Saturn is for middle of month.

MAG   magnitude, the apparent brightness of a planet or star. The lower the number the brighter the object. MAG for Venus, Jupiter and Saturn is for middle of month.

Appendix A: Star Time from Local Time and Date

Date	Evening Hours						Mid-night	Morning Hours					
	6	7	8	9	10	11		1	2	3	4	5	6
Jan 5	1	2	3	4	5	6	7	8	9	10	11	12	13
13	1½	2½	3½	4½	5½	6½	7½	8½	9½	10½	11½	12½	13½
21	2	3	4	5	6	7	8	9	10	11	12	13	14
28	2½	3½	4½	5½	6½	7½	8½	9½	10½	11½	12½	13½	14½
Feb 5	3	4	5	6	7	8	9	10	11	12	13	14	15
13	3½	4½	5½	6½	7½	8½	9½	10½	11½	12½	13½	14½	15½
20	4	5	6	7	8	9	10	11	12	13	14	15	16
28	4½	5½	6½	7½	8½	9½	10½	11½	12½	13½	14½	15½	16½
Mar 7	5	6	7	8	9	10	11	12	13	14	15	16	17
15	5½	6½	7½	8½	9½	10½	11½	12½	13½	14½	15½	16½	17½
22	6	7	8	9	10	11	12	13	14	15	16	17	18
29	6½	7½	8½	9½	10½	11½	12½	13½	14½	15½	16½	17½	18½
Apr 6	7	8	9	10	11	12	13	14	15	16	17	18	19
14	7½	8½	9½	10½	11½	12½	13½	14½	15½	16½	17½	18½	19½
22	8	9	10	11	12	13	14	15	16	17	18	19	20
29	8½	9½	10½	11½	12½	13½	14½	15½	16½	17½	18½	19½	20½
May 7	9	10	11	12	13	14	15	16	17	18	19	20	21
15	9½	10½	11½	12½	13½	14½	15½	16½	17½	18½	19½	20½	21½
22	10	11	12	13	14	15	16	17	18	19	20	21	22
30	10½	11½	12½	13½	14½	15½	16½	17½	18½	19½	20½	21½	22½
Jun 7	11	12	13	14	15	16	17	18	19	20	21	22	23
14	11½	12½	13½	14½	15½	16½	17½	18½	19½	20½	21½	22½	23½
22	12	13	14	15	16	17	18	19	20	21	22	23	24
29	12½	13½	14½	15½	16½	17½	18½	19½	20½	21½	22½	23½	½

Appendix A: Star Time from Local Time and Date (continued)

Date		Evening Hours					Mid-night		Morning Hours				
	6	7	8	9	10	11		1	2	3	4	5	6
Jul 7	13	14	15	16	17	18	19	20	21	22	23	0	1
15	13½	14½	15½	16½	17½	18½	19½	20½	21½	22½	23½	½	1½
22	14	15	16	17	18	19	20	21	22	23	0	1	2
30	14½	15½	16½	17½	18½	19½	20½	21½	22½	23½	½	1½	2½
Aug 6	15	16	17	18	19	20	21	22	23	0	1	2	3
14	15½	16½	17½	18½	19½	20½	21½	22½	23½	½	1½	2½	3½
22	16	17	18	19	20	21	22	23	0	1	2	3	4
29	16½	17½	18½	19½	20½	21½	22½	23½	½	1½	2½	3½	4½
Sep 6	17	18	19	20	21	22	23	0	1	2	3	4	5
13	17½	18½	19½	20½	21½	22½	23½	½	1½	2½	3½	4½	5½
21	18	19	20	21	22	23	0	1	2	3	4	5	6
29	18½	19½	20½	21½	22½	23½	½	1½	2½	3½	4½	5½	6½
Oct 6	19	20	21	22	23	0	1	2	3	4	5	6	7
14	19½	20½	21½	22½	23½	½	1½	2½	3½	4½	5½	6½	7½
21	20	21	22	23	0	1	2	3	4	5	6	7	8
29	20½	21½	22½	23½	½	1½	2½	3½	4½	5½	6½	7½	8½
Nov 6	21	22	23	0	1	2	3	4	5	6	7	8	9
13	21½	22½	23½	½	1½	2½	3½	4½	5½	6½	7½	8½	9½
21	22	23	0	1	2	3	4	5	6	7	8	9	10
28	22½	23½	½	1½	2½	3½	4½	5½	6½	7½	8½	9½	10½
Dec 6	23	0	1	2	3	4	5	6	7	8	9	10	11
14	23½	½	1½	2½	3½	4½	5½	6½	7½	8½	9½	10½	11½
21	0	1	2	3	4	5	6	7	8	9	10	11	12
29	½	1½	2½	3½	4½	5½	6½	7½	8½	9½	10½	11½	12½

Appendix B: Calendar for the Moon and Four Planets

1990

January

Moon	First 4	Full 11	
	Last 18	New 26	
Venus	PM to 12 / AM after 23		

	DAY	RAʰ	DEC°	MAG
Mars	5	16.7	S22	+ 1.6
	15	17.2	S23	+ 1.5
	25	17.8	S24	+ 1.4
Jupiter	15	6.2	N23	− 2.7
Saturn		-o- to 18		
	19	19.3	S22	+ 0.8

February

Moon	First 2	Full 9	
	Last 17	New 25	
Venus	AM		− 4.6

	DAY	RAʰ	DEC°	MAG
Mars	5	18.3	S24	+ 1.4
	15	18.9	S23	+ 1.3
	25	19.4	S23	+ 1.2
Jupiter	15	6.1	N23	− 2.5
Saturn	15	19.5	S22	+ 0.6

March

Moon	First 4	Full 11	
	Last 19	New 26	
Venus	AM		− 4.5

	DAY	RAʰ	DEC°	MAG
Mars	5	19.8	S22	+ 1.2
	15	20.3	S21	+ 1.1
	25	20.8	S19	+ 1.1
Jupiter	15	6.1	N23	− 2.3
Saturn	15	19.7	S21	+ 0.6

April

Moon	First 2	Full 10	
	Last 18	New 25	
Venus	AM		− 4.2

	DAY	RAʰ	DEC°	MAG
Mars	5	21.4	S17	+ 1.0
	15	21.9	S14	+ 0.9
	25	22.4	S12	+ 0.8
Jupiter	15	6.3	N23	− 2.1
Saturn	15	19.8	S21	+ 0.6

May

Moon	First 1	Full 9	
	Last 17	New 24	
	First 31		
Venus	AM		− 4.2

	DAY	RAʰ	DEC°	MAG
Mars	5	22.8	S9	+ 0.8
	15	23.3	S6	+ 0.7
	25	23.8	S4	+ 0.6
Jupiter	15	6.7	N23	− 1.9
Saturn	15	19.8	S21	+ 0.4

June

Moon	Full 8	Last 16	
	New 22	First 29	
Venus	AM		− 4.0

	DAY	RAʰ	DEC°	MAG
Mars	5	0.3	0	+ 0.5
	15	0.4	N2	+ 0.4
	25	1.1	N18	+ 0.4
Jupiter	15	7.1	N23	− 1.9
Saturn	15	19.7	S21	+ 0.2

Nearby reference stars—all magnitude about 1.2: Aldebaran 4.6ʰ N17°,

Appendix B: Calendar for the Moon and Four Planets (continued)

1990

July					October				
Moon	Full 8	Last 15			Moon	Full 4	Last 11		
	New 22	First 29				New 18	First 26		
Venus	AM			− 3.9	Venus		-o-		
	DAY	RA^h	DEC°	MAG		DAY	RA^h	DEC°	MAG
Mars	5	1.6	N8	+ 0.3	Mars	5	4.8	N21	− 1.0
	15	2.0	N10	+ 0.2		15	4.9	N22	− 1.3
	25	2.4	N12	0		25	4.9	N22	− 1.5
Jupiter		-o-			Jupiter	15	8.9	N18	− 2.1
Saturn	15	19.6	S22	+ 0.1	Saturn	15	19.4	S22	+ 0.5

August					November				
Moon	Full 6	Last 13			Moon	Full 2	Last 9		
	New 20	First 28				New 17	First 25		
Venus	AM			− 3.9	Venus		-o-		
	DAY	RA^h	DEC°	MAG		DAY	RA^h	DEC°	MAG
Mars	5	2.9	N14	− 0.1	Mars	5	4.8	N23	− 1.5
	15	3.2	N16	− 0.2		15	4.6	N23	− 1.7
	25	3.7	N18	− 0.3		25	4.3	N23	− 1.9
Jupiter	15	8.1	N21	− 1.8	Jupiter	15	9.1	N17	− 1.8
Saturn	15	19.4	S22	+ 0.2	Saturn	15	19.5	S22	+ 0.8

September					December				
Moon	Full 5	Last 11			Moon	Full 2	Last 9		
	New 19	First 27				New 17	First 25		
Venus	AM to 26			− 3.9		Full 31			
	DAY	RA^h	DEC°	MAG	Venus	PM after 15			− 3.9
Mars	5	4.0	N19	− 0.5		DAY	RA^h	DEC°	MAG
	15	4.3	N20	− 0.6	Mars	5	4.0	N22	− 1.9
	25	4.6	N21	− 0.8		15	3.8	N22	− 1.6
Jupiter	15	8.5	N19	− 1.9		25	3.7	N22	− 1.2
Saturn	15	19.4	S22	+ 0.4	Jupiter	15	9.1	N17	− 2.4
					Saturn	15	19.7	S22	+ 0.2

Pollux 7.8^h N28°, Regulus 10.1^h N12°, Spica 13.4^h S11°, Antares 16.5^h S26°

Appendix B: Calendar for the Moon and Four Planets (continued)

1991

January

Moon	Last	7	New	15	
	First	23	Full	30	
Venus	PM				− 4.5

Mars	DAY	RAʰ	DEC°	MAG
Mars	5	3.7	N22	− 0.9
	15	3.7	N22	− 0.6
	25	3.9	N23	− 0.3
Jupiter	15	8.9	N18	− 2.6
Saturn	-o- after 10			+ 0.5

February

Moon	Last	6	New	14	
	First	21	Full	28	
Venus	PM				− 3.9

Mars	DAY	RAʰ	DEC°	MAG
Mars	5	4.1	N23	0
	15	4.4	N24	+ 0.2
	25	4.6	N24	+ 0.4
Jupiter	15	8.6	N19	− 2.6
Saturn	15	20.2	S20	+ 0.6

March

Moon	Last	8	New	16	
	First	23	Full	30	
Venus	PM				− 3.9

Mars	DAY	RAʰ	DEC°	MAG
Mars	5	4.9	N25	+ 0.6
	15	5.3	N25	+ 0.8
	25	5.7	N25	+ 1.0
Jupiter	15	8.4	N20	− 2.4
Saturn	15	20.4	S20	+ 0.7

April

Moon	Last	7	New	14	
	First	21	Full	28	
Venus	PM				− 4.5

Mars	DAY	RAʰ	DEC°	MAG
Mars	5	6.1	N25	+ 1.1
	15	6.5	N25	+ 1.2
	25	6.9	N25	+ 1.3
Jupiter	15	8.4	N20	− 2.2
Saturn	15	20.6	S19	+ 0.7

May

Moon	Last	7	New	14	
	First	20	Full	28	
Venus	PM				− 4.2

Mars	DAY	RAʰ	DEC°	MAG
Mars	5	7.3	N24	+ 1.4
	15	7.6	N23	+ 1.5
	25	8.1	N22	+ 1.6
Jupiter	15	8.6	N19	− 2.0
Saturn	15	20.6	S19	+ 0.6

June

Moon	Last	5	New	12	
	First	19	Full	27	
Venus	PM				− 4.3

Mars	DAY	RAʰ	DEC°	MAG
Mars	5	8.5	N20	+ 1.6
	15	9.0	N19	+ 1.7
	25	9.3	N17	+ 1.7
Jupiter	15	8.9	N18	− 1.9
Saturn	15	20.6	S19	+ 0.4

Nearby reference stars—all magnitude about 1.2: Aldebaran 4.6ʰ N17°,

Appendix B: Calendar for the Moon and Four Planets (continued)

1991

July

Moon	Last 5	New 11	
	First 18	Full 18	
Venus	PM		− 4.5

	DAY	RA^h	DEC°	MAG

Mars, Jupiter, Saturn:

	DAY	RA	DEC	MAG
Mars	5	9.7	N15	+ 1.8
	15	10.1	N13	+ 1.8
	25	10.5	N10	+ 1.8
Jupiter	15	9.3	N16	− 1.8
Saturn	15	20.2	S21	+ 0.5

August

Moon	Last 3	New 10	
	First 17	Full 25	
Venus	PM to 17 / AM after 27		

	DAY	RA	DEC	MAG
Mars	5	11.0	N8	+ 1.8
	15	11.3	N5	+ 1.8
	25	11.7	N3	+ 1.8
Jupiter		-o- after 4		
Saturn	15	20.3	S20	+ 0.2

September

Moon	Last 1	New 8	
	First 15	Full 23	
Venus	AM		− 4.5

	DAY	RA	DEC	MAG
Mars	5	12.2	0	+ 1.8
	15	12.6	S3	+ 1.7
	25	13.0	S6	+ 1.7
Jupiter	15	10.2	N12	− 1.7
Saturn	15	20.2	S21	+ 0.3

October

Moon	Last 1	New 7	
	First 15	Full 23	
	Last 30		
Venus	PM		− 4.5

	DAY	RA	DEC	MAG
Mars	5		-o- after 3	
	15		-o-	
	25		-o-	
Jupiter	15	10.6	N10	− 1.8
Saturn	15	20.2	S21	+ 0.5

November

Moon	New 6	First 14	
	Full 21	Last 28	
Venus	AM		− 4.3

	DAY	RA	DEC	MAG
Mars	5		-o-	
	15		-o-	
	25		-o-	
Jupiter	15	10.9	N8	− 1.9
Saturn	15	20.3	S20	+ 0.6

December

Moon	New 6	First 14	
	Full 21	Last 28	
Venus	AM		− 4.1

	DAY	RA	DEC	MAG
Mars	5		-o- to 10	
	15	16.6	S22	+ 1.5
	25	17.2	S23	+ 1.5
Jupiter	15	11.0	N7	− 1.7
Saturn	15	20.4	S20	+ 0.7

Pollux 7.8^h N28°, Regulus 10.1^h N12°, Spica 13.4^h S11°, Antares 16.5^h S26°

Appendix B: Calendar for the Moon and Four Planets (continued)

1992

January

Moon	New 4	First 13		
	Full 19	Last 26		
Venus	AM			− 4.0
	DAY	**RA**^h	**DEC**°	**MAG**
Mars	5	17.8	S24	+ 1.4
	15	18.3	S24	+ 1.4
	25	18.8	S24	+ 1.4
Jupiter	15	11.1	N8	− 2.3
Saturn	15	20.7	S18	+ 0.8
	-o- after 18			

February

Moon	New 3	First 11		
	Full 18	Last 25		
Venus	AM			− 4.0
	DAY	**RA**^h	**DEC**°	**MAG**
Mars	5	19.4	S23	+ 1.3
	15	20.0	S23	+ 1.3
	25	20.5	S20	+ 1.3
Jupiter	15	10.9	N8	− 2.5
Saturn		-o- to 10		
	15	20.9	S18	+ 0.7

March

Moon	New 4	First 12		
	Full 18	Last 26		
Venus	AM			− 3.9
	DAY	**RA**^h	**DEC**°	**MAG**
Mars	5	21.0	S18	+ 1.3
	15	21.5	S16	+ 1.2
	25	22.0	S13	+ 1.2
Jupiter	15	10.7	N10	− 2.5
Saturn	15	21.1	S17	+ 0.8

April

Moon	New 3	First 10		
	Full 17	Last 24		
Venus	AM			− 3.9
	DAY	**RA**^h	**DEC**°	**MAG**
Mars	5	22.2	S9	1.1
	15	23.0	S6	1.1
	25	23.5	S5	1.1
Jupiter	15	10.5	N11	− 2.3
Saturn	15	21.3	S16	+ 0.8

May

Moon	New 2	First 9		
	Full 18	Last 24		
Venus	-o- after 8			− 3.9
	DAY	**RA**^h	**DEC**°	**MAG**
Mars	5	0.0	S2	+ 1.1
	15	0.5	N2	+ 1.0
	25	0.9	N5	+ 1.0
Jupiter	15	10.5	N11	− 2.0
Saturn	15	21.4	S16	+ 0.7

June

Moon	New 1	First 7		
	Full 15	Last 23		
	New 30			
Venus		-o-		
	DAY	**RA**^h	**DEC**°	**MAG**
Mars	5	1.4	N8	+ 1.0
	15	1.9	N10	+ 0.9
	25	2.4	N13	+ 0.9
Jupiter	15	10.6	N10	− 1.9
Saturn	15	21.4	S16	+ 0.5

Nearby reference stars—all magnitude about 1.2: Aldebaran 4.6^h N17°,

Appendix B: Calendar for the Moon and Four Planets (continued)

1992

July				
Moon	First 7	Full 14		
	Last 22	New 29		
Venus	PM after 22			− 3.9
	DAY	**RAʰ**	**DEC°**	**MAG**
Mars	5	2.8	N15	+ 0.9
	15	3.3	N17	+ 0.8
	25	3.8	N19	+ 0.8
Jupiter	15	10.9	N8	− 1.8
Saturn	15	21.3	S17	+ 0.5

August				
Moon	First 5	Full 13		
	Last 21	New 28		
Venus	PM			− 3.9
	DAY	**RAʰ**	**DEC°**	**MAG**
Mars	5	4.3	N21	+ 0.7
	15	4.8	N22	+ 0.7
	25	5.2	N23	+ 0.6
Jupiter	15	11.3	N6	− 1.7
Saturn	15	21.0	S18	+ 0.5

September				
Moon	First 3	Full 12		
	Last 19	New 26		
Venus	PM			− 3.9
	DAY	**RAʰ**	**DEC°**	**MAG**
Mars	5	5.7	N23	− 0.6
	15	6.1	N23	+ 0.5
	25	6.5	N23	+ 0.4
Jupiter			-o- after 5	
Saturn	15	21.0	S18	+ 0.5

October				
Moon	First 3	Full 11		
	Last 19	New 25		
Venus	PM			− 4.0
	DAY	**RAʰ**	**DEC°**	**MAG**
Mars	5	6.9	N23	+ 0.3
	15	7.2	N23	+ 0.2
	25	7.5	N23	+ 0.0
Jupiter	15	12.1	N1	− 1.7
Saturn	15	21.0	S18	+ 0.6

November				
Moon	First 2	Full 9		
	Last 17	New 24		
Venus	PM			4.1
	DAY	**RAʰ**	**DEC°**	**MAG**
Mars	5	8.0	N23	− 0.8
	15	7.9	N24	− 1.1
	25	7.7	N25	− 1.3
Jupiter	15	12.5	S2	− 1.9
Saturn	15	21.1	S17	+ 0.7

December				
Moon	First 2	Full 9		
	Last 16	New 24		
Venus	PM			− 4.2
	DAY	**RAʰ**	**DEC°**	**MAG**
Mars	5	8.0	N23	− 0.8
	15	7.9	N24	− 1.1
	25	7.7	N25	− 1.3
Jupiter	15	12.7	S3	− 1.9
Saturn	15	22.1	S17	+ 0.7

Pollux 7.8ʰ N28°, Regulus 10.1ʰ N12°, Spica 13.4ʰ S11°, Antares 16.5ʰ S26°

Appendix B: Calendar for the Moon and Four Planets (continued)

1993

January

Moon	First	1	Full	8	
	Last	15	New	22	
	First	30			
Venus	PM				− 4.4

	DAY	RA^h	DEC°	MAG
Mars	5	7.4	N26	− 1.4
	15	7.1	N27	− 1.3
	25	6.9	N27	− 1.1
Jupiter	15	12.9	S4	− 2.1
Saturn	15	21.4	S16	+ 0.7

February

Moon	Full	6	Last	13	
	New	21			
Venus	PM				− 4.4

	DAY	RA^h	DEC°	MAG
Mars	5	6.7	N27	− 0.9
	15	6.6	N27	− 0.6
	25	6.7	N26	− 0.3
Jupiter	15	12.9	S4	− 2.3
Saturn	-o- to 27			+ 0.9

March

Moon	First	1	Full	8	
	Last	15	New	23	
	First	31			
Venus	PM				− 4.4

	DAY	RA^h	DEC°	MAG
Mars	5	6.8	N26	+ 0.0
	15	6.9	N26	+ 0.2
	25	7.2	N25	+ 0.4
Jupiter	15	12.8	S3	− 2.4
Saturn	15	21.8	S14	+ 0.8

April

Moon	Full	6	Last	13	
	New	21	First	29	
Venus	AM after 6				− 4.3

	DAY	RA^h	DEC°	MAG
Mars	5	7.5	N24	+ 0.6
	15	7.8	N23	+ 0.8
	25	8.1	N22	+ 0.9
Jupiter	15	12.5	S2	− 2.4
Saturn	15	22.0	S13	+ 0.9

May

Moon	Full	6	Last	13	
	New	21	First	28	
Venus	AM				− 4.5

	DAY	RA^h	DEC°	MAG
Mars	5	8.4	N21	+ 1.1
	15	8.7	N20	+ 1.2
	25	9.1	N18	+ 1.3
Jupiter	15	12.3	N1	− 2.3
Saturn	15	22.2	S13	+ 0.9

June

Moon	Full	4	Last	12	
	New	20	First	26	
Venus	AM				− 4.3

	DAY	RA^h	DEC°	MAG
Mars	5	9.5	N16	+ 1.4
	15	9.9	N14	+ 1.4
	25	10.2	N12	+ 1.5
Jupiter	15	12.3	N1	− 2.1
Saturn	15	22.2	S13	+ 0.8

Nearby reference stars—all magnitude about 1.2: Aldebaran 4.6^h N17°,

Appendix B: Calendar for the Moon and Four Planets (continued)

1993

July

Moon	Full	3	Last	11	
	New	19	First	26	

Venus	AM				− 4.1
	DAY	**RAʰ**	**DEC°**		**MAG**
Mars	5	10.6	N8		+ 1.6
	15	11.0	N10		+ 1.6
	25	11.3	N5		+ 1.6
Jupiter	15	12.5	S2		− 1.9
Saturn	15	22.1	S13		+ 0.6

August

Moon	Full	2	Last	10	
	New	17	First	24	

Venus	AM				− 4.0
	DAY	**RAʰ**	**DEC°**		**MAG**
Mars	5	11.7	N2		+ 1.6
	15	12.1	0		+ 1.7
	25	12.5	S3		+ 1.7
Jupiter	15	12.8	S4		− 1.8
Saturn	15	22.0	S14		+ 0.4

September

Moon	Full	1	Last	9	
	New	16	First	22	
	Full	30			

Venus	AM				− 4.0
	DAY	**RAʰ**	**DEC°**		**MAG**
Mars	5	12.9	S6		+ 1.6
	15	13.3	S8		+ 1.6
	25	13.8	S11		+ 1.6
Jupiter	15	13.1	S6		− 1.7
Saturn	15	21.8	S15		+ 0.4

October

Moon	Last	8	New	15	
	First	22	Full	30	

Venus	AM				− 4.0
	DAY	**RAʰ**	**DEC°**		**MAG**
Mars	5	14.2	S13		+ 1.6
	15	14.6	S16		+ 1.5
	25	15.1	S18		+ 1.5
Jupiter		-o- after 5			
Saturn	15	21.7	S15		+ 0.5

November

Moon	Last	7	New	13	
	First	21	Full	29	

Venus	AM				− 3.9
	DAY	**RAʰ**	**DEC°**		**MAG**
Mars	5	15.6	S20		+ 1.5
	15	16.1	S21		+ 1.4
		-o- after 19			
Jupiter	15	14.0	S11		− 1.7
Saturn	15	21.8	S15		+ 0.7

December

Moon	Last	6	New	13	
	First	20	Full	28	
Venus		-o- after 4			

	DAY	**RAʰ**	**DEC°**		**MAG**
Mars		-o- all month			
Jupiter	15	14.3	S13		− 1.7
Saturn	15	21.9	S14		+ 0.8

Pollux 7.8ʰ N28°, Regulus 10.1ʰ N12°, Spica 13.4ʰ S11°, Antares 16.5ʰ S26°

Appendix B: Calendar for the Moon and Four Planets (continued)

1994

January

Moon	Last 5	New 11		
	First 19	Full 27		
Venus		-o-		
	DAY	**RA^h**	**DEC°**	**MAG**
Mars		-o- all month		
Jupiter	15	14.6	S14	− 1.9
Saturn	15	22.0	S13	+ 0.9

February

Moon	Last 3	New 10		
	First 18	Full 26		
Venus	PM after 27			− 3.9
	DAY	**RA^h**	**DEC°**	**MAG**
Mars		-o- to 7		
	15	21.1	S18	+ 1.2
	25	21.6	S15	+ 1.2
Jupiter	15	14.8	S15	− 2.1
Saturn		-o- after 4		

March

Moon	Last 4	New 12		
	First 20	Full 27		
Venus	PM			− 3.9
	DAY	**RA^h**	**DEC°**	**MAG**
Mars	5	22.0	S13	+ 1.2
	15	22.5	S10	+ 1.2
	25	23.0	S7	+ 1.2
Jupiter	15	14.8	S15	− 2.1
Saturn		-o- to 6		
	15	22.5	S11	+ 0.9

April

Moon	Last 3	New 11		
	First 19	Full 25		
Venus	PM			− 3.9
	DAY	**RA^h**	**DEC°**	**MAG**
Mars	5	23.6	S4	+ 1.2
	15	0.0	N1	+ 1.2
	25	0.5	N2	+ 1.2
Jupiter	15	14.6	S14	− 2.5
Saturn	15	22.7	S10	+ 1.1

May

Moon	Last 2	New 10		
	First 18	Full 25		
Venus	PM			− 4.0
	DAY	**RA^h**	**DEC°**	**MAG**
Mars	5	1.0	N5	+ 1.2
	15	1.5	N8	+ 1.2
	25	2.0	N11	+ 1.2
Jupiter	15	14.4	S13	− 2.5
Saturn	15	22.9	S9	+ 1.1

June

June	Last 1	New 9		
	First 16	Full 23		
	Last 30			
Venus	PM			− 4.0
	DAY	**RA^h**	**DEC°**	**MAG**
Mars	5	2.4	N14	+ 1.2
	15	2.9	N16	+ 1.2
	25	3.4	N18	+ 1.2
Jupiter	15	14.2	S12	− 2.3
Saturn	15	22.9	S9	+ 1.0

Nearby reference stars—all magnitude about 1.2: Aldebaran 4.6^h N17°,

Appendix B: Calendar for the Moon and Four Planets (continued)

1994

July

Moon	New 8	First 16		
	Full 22	Last 30		
Venus	PM			− 4.1

Mars	DAY	RAʰ	DEC°	MAG
Mars	5	3.9	N20	+ 1.2
	15	4.4	N21	+ 1.2
	25	4.9	N22	+ 1.2
Jupiter	15	14.2	S12	− 2.1
Saturn	15	22.9	S9	+ 0.8

August

Moon	New 7	First 14		
	Full 21	Last 29		
Venus	PM			− 4.3

Mars	DAY	RAʰ	DEC°	MAG
Mars	5	5.4	N23	+ 1.2
	15	5.9	N24	+ 1.2
	25	6.4	N24	+ 1.2
Jupiter	15	14.4	S13	− 1.9
Saturn	15	22.8	S10	+ 0.6

September

Moon	New 5	First 12		
	Full 19	Last 28		
Venus	PM			− 4.5

Mars	DAY	RAʰ	DEC°	MAG
Mars	5	6.9	N23	+ 1.1
	15	7.3	N23	+ 1.1
	25	7.8	N22	+ 1.1
Jupiter	15	14.7	S15	− 1.8
Saturn	15	22.6	S11	+ 0.5

October

Moon	New 5	First 11		
	Full 19	Last 27		
Venus	PM to 27			− 4.5

Mars	DAY	RAʰ	DEC°	MAG
Mars	5	8.2	N21	+ 1.0
	15	8.6	N20	+ 0.9
	25	8.9	N18	+ 0.8
Jupiter	15	15.0	S16	− 1.7
Saturn	15	22.6	S11	+ 0.7

November

Moon	New 3	First 10		
	Full 18	Last 26		
Venus	AM after 9			− 4.5

Mars	DAY	RAʰ	DEC°	MAG
Mars	5	9.3	N18	+ 0.7
	15	9.6	N16	+ 0.6
	25	9.9	N16	+ 0.4
Jupiter		-o- after 4		
Saturn	15	22.5	S11	+ 0.8

December

Moon	New 2	First 9		
	Full 18	Last 25		
Venus	AM			4.6

Mars	DAY	RAʰ	DEC°	MAG
Mars	5	10.1	N15	+ 0.2
	15	10.2	N14	+ 0.1
	25	10.3	N14	− 0.2
Jupiter	15	16.0	S20	− 1.7
Saturn	15	22.6	S11	+ 0.9

Pollux 7.8ʰ N28°, Regulus 10.1ʰ N12°, Spica 13.4ʰ S11°, Antares 16.5ʰ S26°

Appendix B: Calendar for the Moon and Four Planets (continued)

1995

January

Moon	New 1		First 8	
	Full 16		Last 24	
	New 30			
Venus	AM			− 4.5
	DAY	**RA**ʰ	**DEC°**	**MAG**
Mars	5	10.4	N14	− 0.4
	15	10.3	N15	− 0.6
	25	10.2	N16	− 0.9
Jupiter	15	16.4	S21	− 1.8
Saturn	15	22.7	S10	+ 1.0

February

Moon	First 7		Full 15	
	Last 22			
Venus	AM			− 4.2
	DAY	**RA**ʰ	**DEC°**	**MAG**
Mars	5	10.0	N17	− 1.1
	15	9.7	N19	− 1.2
	25	9.4	N20	− 1.1
Jupiter	15	16.7	S22	− 2.0
Saturn	15	23.0	S9	+ 1.0
		-o- after 23		

March

Moon	New 1		First 9	
	Full 17		Last 23	
	New 31			
Venus	AM			− 4.0
	DAY	**RA**ʰ	**DEC°**	**MAG**
Mars	5	9.3	N20	− 0.9
	15	9.1	N20	− 0.6
	25	9.1	N20	− 0.4
Jupiter	15	16.9	S22	− 2.1
Saturn		-o- to 18		

April

Moon	First 8		Full 15	
	Last 22		New 29	
Venus	AM			− 3.9
	DAY	**RA**ʰ	**DEC°**	**MAG**
Mars	5	9.1	N20	− 0.1
	15	9.3	N19	+ 0.1
	25	9.4	N18	+ 0.3
Jupiter	15	16.9	S22	− 2.4
Saturn	15	23.4	S6	+ 1.2

May

Moon	First 7		Full 14	
	Last 22		New 29	
Venus	AM			− 3.9
	DAY	**RA**ʰ	**DEC°**	**MAG**
Mars	5	9.6	N16	+ 0.5
	15	9.9	N15	+ 0.7
	25	10.1	N13	+ 0.8
Jupiter	15	16.7	S21	− 2.5
Saturn	15	23.5	S5	+ 1.3

June

Moon	First 6		Full 13	
	Last 19		New 28	
Venus	AM			− 3.9
	DAY	**RA**ʰ	**DEC°**	**MAG**
Mars	5	10.5	N11	+ 0.9
	15	10.8	N9	+ 1.0
	25	11.1	N7	+ 1.1
Jupiter	15	16.5	S21	− 2.6
Saturn	15	23.7	S4	+ 1.2

Nearby reference stars—all magnitude about 1.2: Aldebaran 4.6ʰ N17°,

Appendix B: Calendar for the Moon and Four Planets (continued)

1995

July

Moon	First	5	Full	12	
	Last	19	New	27	

Venus	AM to 17			− 3.9
	DAY	**RA**ʰ	**DEC**°	**MAG**
Mars	5	11.4	N4	+ 1.2
	15	11.8	N2	+ 1.3
	25	12.1	0	+ 1.3
Jupiter	15	16.3	S21	− 2.4
Saturn	15	23.7	S4	+ 1.0

August

Moon	First	4	Full	10
	Last	18	New	26

Venus		-o-		
	DAY	**RA**ʰ	**DEC**°	**MAG**
Mars	5	12.5	S3	+ 1.3
	15	12.9	S6	+ 1.4
	25	13.3	S8	+ 1.4
Jupiter	15	16.3	S21	− 2.3
Saturn	15	23.6	S5	+ 0.9

September

Moon	First	2	Full	9
	Last	16	New	24

Venus	PM after 25			− 3.9
	DAY	**RA**ʰ	**DEC**°	**MAG**
Mars	5	13.8	S11	+ 1.4
	15	14.2	S14	+ 1.4
	25	14.6	S16	+ 1.4
Jupiter	15	16.4	S21	− 2.0
Saturn	15	23.5	S6	+ 0.7

October

Moon	First	1	Full	8
	Last	16	New	24
	First	30		

Venus	PM			− 3.9
	DAY	**RA**ʰ	**DEC**°	**MAG**
Mars	5	15.1	S18	+ 1.4
	15	15.5	S20	+ 1.4
	25	16.0	S21	+ 1.4
Jupiter	15	16.8	S22	− 2.0
Saturn	15	23.4	S7	+ 0.7

November

Moon	Full	7	Last	15
	New	22	First	29

Venus	PM			3.9
	DAY	**RA**ʰ	**DEC**°	**MAG**
Mars	5	16.6	S23	+ 1.3
	15	17.1	S24	+ 1.3
	25	17.7	S24	+ 1.3
Jupiter	15	17.2	S23	− 1.8
Saturn	15	23.3	S7	+ 0.9

December

Moon	Full	7	Last	15
	New	22	First	28

Venus	PM			− 4.0
	DAY	**RA**ʰ	**DEC**°	**MAG**
Mars	5	18.2	S24	+ 1.2
	15	18.8	S24	+ 1.2
	25	19.4	S23	+ 1.2
Jupiter		-o- after 5		
Saturn	15	23.3	S7	+ 1.1

Pollux 7.8ʰ N28°, Regulus 10.1ʰ N12°, Spica 13.4ʰ S11°, Antares 16.5ʰ S26°

Appendix B: Calendar for the Moon and Four Planets (continued)

1996

January

Moon	Full	5	Last	13
	New 20		First 27	

Venus PM   − 4.0

	DAY	RA^h	DEC°	MAG
Mars	5	20.0	S22	+ 1.2
	15	20.5	S20	+ 1.2
	-o- after 18			
Jupiter	15	18.2	S23	− 1.9
Saturn	15	23.4	S6	+ 1.1

February

Moon	Full	4	Last	12
	New 18		First 26	

Venus PM   − 4.1

	DAY	RA^h	DEC°	MAG
Mars	-o- all month			
Jupiter	15	18.6	S2	− 1.9
Saturn	15	23.6	S5	+ 1.2

March

Moon	Full	5	Last	12
	New 19		First 27	

Venus PM   − 4.2

	DAY	RA^h	DEC°	MAG
Mars	-o- all month			
Jupiter	15	19.0	S23	− 2.1
Saturn	-o- after 6			

April

Moon	Full	4	Last	10
	New 17		First 25	

Venus PM   − 4.4

	DAY	RA^h	DEC°	MAG
Mars	-o- to 27			+ 1.1
Jupiter	15	19.2	S22	− 2.3
Saturn	15	0.1	S2	+ 1.1

May

Moon	Full	3	Last	10
	New 17		First 25	

Venus PM   − 4.5

	DAY	RA^h	DEC°	MAG
Mars	5	2.0	N12	+ 1.3
	15	2.5	N14	+ 1.3
	25	3.0	N17	+ 1.4
Jupiter	15	19.3	S22	− 2.5
Saturn	15	0.3	0	+ 1.0

June

Moon	Full	1	Last	8
	New 16		First 24	

Venus AM after 18   − 4.0

	DAY	RA^h	DEC°	MAG
Mars	5	3.5	N19	+ 1.4
	15	4.0	N20	+ 1.4
	25	4.5	N22	+ 1.4
Jupiter	15	19.1	S23	− 2.6
Saturn	15	0.4	0	+ 1.0

Nearby reference stars—all magnitude about 1.2: Aldebaran 4.6^h N17°,

Appendix B: Calendar for the Moon and Four Planets (continued)

1996

July

Moon	Full	1	Last	7	
	New 15		First 23		

Venus	AM				− 4.5
	DAY	RA^h	DEC°	MAG	
Mars	5	5.0	N23	+ 1.5	
	15	5.5	N24	+ 1.5	
	25	6.0	N24	+ 1.5	
Jupiter	15	18.8	S23	− 2.7	
Saturn	15	0.5	N1	+ 0.8	

August

Moon	Last	6	New 14	
	First 22		Full 28	

Venus	AM				− 4.5
	DAY	RA^h	DEC°	MAG	
Mars	5	6.5	N24	+ 1.5	
	15	7.0	N24	+ 1.5	
	25	7.5	N22	+ 1.5	
Jupiter	15	18.6	S23	− 2.6	
Saturn	15	0.5	0	+ 0.6	

September

Moon	Last	4	New 12	
	First 20		Full 27	

Venus	AM				− 4.2
	DAY	RA^h	DEC°	MAG	
Mars	5	8.0	N22	+ 1.5	
	15	8.4	N20	+ 1.5	
	25	8.8	N19	+ 1.5	
Jupiter	15	18.6	S23	− 2.3	
Saturn	15	0.3	0	+ 0.5	

October

Moon	Last	4	New 12	
	First 19		Full 26	

Venus	AM				− 4.1
	DAY	RA^h	DEC°	MAG	
Mars	5	9.2	N17	+ 1.4	
	15	9.6	N16	+ 1.4	
	25	10.0	N14	+ 1.3	
Jupiter	15	18.7	S23	− 2.2	
Saturn	15	0.2	S1	+ 0.6	

November

Moon	Last	3	New 11	
	First 18		Full 25	

Venus	AM				− 4.0
	DAY	RA^h	DEC°	MAG	
Mars	5	10.4	N12	+ 1.2	
	15	10.7	N10	+ 1.2	
	25	11.0	N8	+ 1.1	
Jupiter	15	19.1	S23	− 2.0	
Saturn	15	0.1	S2	+ 0.8	

December

Moon	Last	3	New 10	
	First 17		Full 24	

Venus	AM				− 4.0
	DAY	RA^h	DEC°	MAG	
Mars	5	11.3	N7	+ 0.9	
	15	11.6	N5	+ 0.8	
	25	11.9	N3	+ 0.6	
Jupiter	15	19.5	S22	− 1.9	
Saturn	15	0.1	S2	+ 0.9	

Pollux 7.8^h N28°, Regulus 10.1^h N12°, Spica 13.4^h S11°, Antares 16.5^h S26°

Appendix B: Calendar for the Moon and Four Planets (continued)

1997

January

| Moon | Last | 2 | New | 9 | |
| | First | 15 | Full | 23 | |

| Venus | AM | | | | − 3.9 |

	DAY	RA^h	DEC°	MAG
Mars	5	12.1	N2	+ 0.5
	15	12.2	N1	+ 0.2
	25	12.4	N1	0.0
Jupiter		-o- after 6		
Saturn	15	0.2	S1	+ 1.0

February

| Moon | New | 7 | First | 14 |
| | Full | 22 | | |

| Venus | AM to 20 | | | − 3.9 |

	DAY	RA^h	DEC°	MAG
Mars	5	12.4	N1	− 0.2
	15	12.4	N1	− 0.5
	25	12.3	N2	− 0.8
Jupiter	15	20.5	S19	− 1.9
Saturn	15	0.3	0	+ 1.0

March

Moon	Last	2	New	9
	First	16	Full	24
	Last	31		

| Venus | | -o- | | |

	DAY	RA^h	DEC°	MAG
Mars	5	12.2	N3	− 1.0
	15	12.0	N4	− 1.2
	25	11.7	N6	− 1.2
Jupiter	15	21.0	S18	− 2.0
Saturn	15	0.5	N1	+ 0.9
	-o- after 19			

April

| Moon | New | 7 | First | 14 |
| | Full | 22 | Last | 30 |

| Venus | | -o- | | |

	DAY	RA^h	DEC°	MAG
Mars	5	11.5	N7	− 1.1
	15	11.3	N7	− 0.9
	25	11.3	N7	− 0.6
Jupiter	15	21.3	S16	− 2.1
Saturn		-o- to 12		
	15	0.8	N3	+ 0.8

May

| Moon | New | 6 | First | 14 |
| | Full | 22 | Last | 29 |

| Venus | PM after 14 | | | − 3.9 |

	DAY	RA^h	DEC°	MAG
Mars	5	11.2	N7	0.4
	15	11.3	N6	− 0.2
	25	11.5	N5	0.0
Jupiter	15	21.6	S15	− 2.3
Saturn	15	1.0	N4	+ 0.8

June

| Moon | New | 5 | First | 13 |
| | Full | 20 | Last | 27 |

| Venus | PM | | | − 3.9 |

	DAY	RA^h	DEC°	MAG
Mars	5	11.7	N3	+ 0.2
	15	11.9	N1	+ 0.3
	25	12.2	N1	+ 0.5
Jupiter	15	21.6	S15	− 2.5
Saturn	15	1.2	N5	+ 0.7

Pollux 7.8^h N28°, Regulus 10.1^h N12°, Spica 13.4^h S11°, Antares 16.5^h S26°

Appendix B: Calendar for the Moon and Four Planets (continued)

1997

July

Moon	New 4	First 12		
	Full 20	Last 26		
Venus	PM			− 3.9

Mars	DAY	RA^h	DEC°	MAG
Mars	5	12.4	S3	+ 0.3
	15	12.8	S5	+ 0.7
	25	13.1	S7	+ 0.8
Jupiter	15	21.5	S16	− 2.7
Saturn	15	1.3	N5	+ 0.6

August

Moon	New 3	First 11		
	Full 18	Last 25		
Venus	PM			− 4.0

Mars	DAY	RA^h	DEC°	MAG
Mars	5	13.5	S10	+ 0.8
	15	13.9	S12	+ 0.9
	25	14.3	S14	+ 1.0
Jupiter	15	21.3	S17	− 2.8
Saturn	15	1.3	N5	+ 0.4

September

Moon	New 1	First 10		
	Full 16	Last 23		
Venus	PM			− 4.1

Mars	DAY	RA^h	DEC°	MAG
Mars	5	14.7	S17	+ 1.0
	15	15.2	S19	+ 1.1
	25	15.6	S20	+ 1.1
Jupiter	15	21.0	S18	− 2.7
Saturn	15	1.2	N5	+ 0.3

October

Moon	New 1	First 9		
	Full 16	Last 23		
	New 31			
Venus	PM			− 4.3

Mars	DAY	RA^h	DEC°	MAG
Mars	5	16.1	S22	+ 1.1
	15	16.6	S23	+ 1.1
	25	17.2	S24	+ 1.1
Jupiter	15	21.0	S18	− 2.5
Saturn	15	1.1	N4	+ 0.2

November

Moon	First 7	Full 14		
	Last 21	New 30		
Venus	PM			− 4.5

Mars	DAY	RA^h	DEC°	MAG
Mars	5	17.8	S25	+ 1.1
	15	18.3	S25	+ 1.1
	25	18.9	S24	+ 1.1
Jupiter	15	21.1	S17	− 2.3
Saturn	15	0.9	N3	+ 0.4

December

Moon	First 7	Full 14		
	Last 21	New 29		
Venus	PM			− 4.7

Mars	DAY	RA^h	DEC°	MAG
Mars	5	19.4	S23	+ 1.2
	15	20.0	S22	+ 1.2
	25	20.6	S20	+ 1.2
Jupiter	15	21.4	S16	− 2.1
Saturn	15	0.9	N3	+ 0.6

Pollux 7.8^h N28°, Regulus 10.1^h N12°, Spica 13.4^h S11°, Antares 16.5^h S26°

Appendix B: Calendar for the Moon and Four Planets (continued)

1998

January

| Moon | First 5 | Full 12 |
| Last 20 | New 28 |

Venus PM to 11
AM after 22

Mars	DAY	RAʰ	DEC°	MAG
Mars	5	21.0	S18	+ 1.2
	15	21.7	S16	+ 1.2
	25	21.1	S13	+ 1.2
Jupiter	15	21.8	S14	− 2.0
Saturn	15	0.9	N3	+ 0.7

February

Moon First 3   Full 11
Last 19   New 26

Venus AM   − 4.6

	DAY	RAʰ	DEC°	MAG
Mars	5	22.7	S9	+ 1.2
	15	23.2	S6	+ 1.2
	25	23.6	S3	+ 1.2
Jupiter		-o- to 9		
Saturn	15	1.1	N4	+ 0.7

March

Moon First 5   Full 13
Last 21   New 28

Venus AM   − 4.5

	DAY	RAʰ	DEC°	MAG
Mars	5	0.0	0	+ 1.2
	15	0.5	N3	+ 1.2
	25	0.6	N6	+ 1.3
		-o- after 27		
Jupiter		-o- to 9		
	15	22.7	S7	− 1.5
Saturn	15	1.2	N6	+ 0.7

April

Moon First 3   Full 11
Last 19   New 26

Venus AM   − 4.2

	DAY	RAʰ	DEC°	MAG
Mars		-o- all month		
Jupiter	15	23.3	S6	− 2.1
Saturn		-o- to 26		

May

Moon First 3   Full 11
Last 19   New 25

Venus AM   − 4.0

	DAY	RAʰ	DEC°	MAG
Mars		-o- all month		
Jupiter	15	23.5	S4	− 2.2
Saturn	15	1.7	N8	+ 0.6

June

Moon First 2   Full 10
Last 17   New 24

Venus AM   − 3.9

	DAY	RAʰ	DEC°	MAG
Mars		-o- to 25		
	25	5.5	N24	+ 1.7
Jupiter	15	23.8	S3	− 2.4
Saturn	15	2.0	N9	+ 0.5

Nearby reference stars—all magnitude about 1.2: Aldebaran 4.6ʰ N17°,

Appendix B: Calendar for the Moon and Four Planets (continued)

1998

July

Moon	First	1	Full	9	
	Last	16	New	23	
Venus	AM				− 3.9

Mars	DAY	RAʰ	DEC°	MAG
Mars	5	5.9	N24	+ 1.6
	15	6.4	N24	+ 1.6
	25	6.9	N24	+ 1.6
Jupiter	15	23.9	S2	− 2.6
Saturn	15	2.1	N10	+ 0.5

August

Moon	Full	8	Last	14
	New	22	Last	30
Venus	AM			− 3.9

Mars	DAY	RAʰ	DEC°	MAG
Mars	5	7.4	N23	+ 1.7
	15	7.9	N22	+ 1.7
	25	8.4	N21	+ 1.7
Jupiter	15	23.8	S3	− 2.8
Saturn	15	2.1	N10	+ 0.4

September

Moon	Full	6	Last	13
	New	20	First	28
Venus	AM to 19			− 3.9

Mars	DAY	RAʰ	DEC°	MAG
Mars	5	8.8	N19	+ 1.7
	15	9.2	N17	+ 1.7
	25	9.6	N15	+ 1.7
Jupiter	15	23.6	S4	− 2.8
Saturn	15	2.1	N10	+ 0.3

October

Moon	Full	5	Last	12
	New	20	First	28
Venus		-o-		

Mars	DAY	RAʰ	DEC°	MAG
Mars	5	10.0	N13	+ 1.7
	15	10.5	N11	+ 1.7
	25	10.8	N9	+ 1.6
Jupiter	15	23.4	S6	− 2.8
Saturn	15	2.0	N9	+ 0.0

November

Moon	Full	4	Last	11
	New	18	First	26
Venus		-o-		

Mars	DAY	RAʰ	DEC°	MAG
Mars	5	11.2	N7	+ 1.6
	15	11.6	N4	+ 1.5
	25	12.0	N2	+ 1.4
Jupiter	15	23.3	S6	− 2.6
Saturn	15	1.8	N8	+ 0.1

December

Moon	Full	3	Last	10
	New	18	First	26
Venus	PM after 12			− 3.9

Mars	DAY	RAʰ	DEC°	MAG
Mars	5	12.3	0	+ 1.4
	15	12.6	S2	+ 1.3
	25	13.0	S4	+ 1.1
Jupiter	15	23.4	S5	− 2.4
Saturn	15	1.7	N8	+ 0.3

Pollux 7.8ʰ N28°, Regulus 10.1ʰ N12°, Spica 13.4ʰ S11°, Antares 16.5ʰ S26°

Appendix B: Calendar for the Moon and Four Planets (continued)

1999

January

Moon	Full	2	Last	9		
	New	17	First	24		
	Full	31				
Venus	PM					− 3.9
	DAY	**RAʰ**	**DEC°**	**MAG**		
Mars	5	13.3	S6	+ 1.0		
	15	13.6	S8	+ 0.9		
	25	13.9	S9	+ 0.7		
Jupiter	15	23.7	S3	− 2.2		
Saturn	15	1.7	N8	+ 0.4		

February

Moon	Last	8	New	16
	First	23		
Venus	PM			− 3.9
	DAY	**RAʰ**	**DEC°**	**MAG**
Mars	5	14.1	S11	+ 0.4
	15	14.3	S12	+ 0.2
	25	14.5	S13	+ 0.0
Jupiter	15	0.0	N1	− 2.1
Saturn	15	1.8	N9	+ 0.5

March

Moon	Full	2	Last	10
	New	17	First	24
	Full	31		
Venus	PM			− 4.0
	DAY	**RAʰ**	**DEC°**	**MAG**
Mars	5	14.6	S13	− 0.2
	15	14.7	S13	− 0.5
	25	14.7	S13	− 0.7
Jupiter	15	0.5	N2	− 2.1
		-o- after 18		
Saturn	15	2.0	N10	+ 0.5

April

Moon	Last	9	New	16
	First	22	Full	30
Venus	AM			− 4.1
	DAY	**RAʰ**	**DEC°**	**MAG**
Mars	5	14.6	S13	− 1.2
	15	14.4	S12	− 1.5
	25	14.0	S12	− 1.6
Jupiter		-o- to 15		
	15	2.2	N4	+ 0.3
Saturn	15	2.2	N11	+ 0.3
		-o- after 16		

May

Moon	Last	8	New	15
	First	22	Full	30
Venus	PM			− 4.2
	DAY	**RAʰ**	**DEC°**	**MAG**
Mars	5	13.9	S11	− 1.6
	15	13.7	S10	− 1.5
	25	13.5	S10	− 1.3
Jupiter	15	1.3	N7	− 2.1
Saturn		-o- to 8		
	15	2.5	N12	+ 0.4

June

Moon	Last	7	New	13
	First	20	Full	28
Venus	PM			− 4.3
	DAY	**RAʰ**	**DEC°**	**MAG**
Mars	5	13.5	S10	− 0.9
	15	13.5	S10	− 0.8
	25	13.6	S11	− 0.6
Jupiter	15	1.7	N10	− 2.2
Saturn	15	2.7	N13	+ 0.4

Nearby reference stars—all magnitude about 1.2: Aldebaran 4.6ʰ N17°,

Appendix B: Calendar for the Moon and Four Planets (continued)

1999

July

Moon	Last 6	New 13		
	First 20	Full 28		
Venus	PM			− 4.5
	DAY	**RAʰ**	**DEC°**	**MAG**
Mars	5	13.8	S13	− 0.3
	15	14.1	S14	− 0.2
	25	14.3	S16	− 0.1
Jupiter	15	2.0	N11	− 2.3
Saturn	15	2.9	N14	+ 0.3

August

Moon	Last 4	New 11		
	First 19	Full 26		
Venus	PM to 15			− 4.0
	AM after 25			− 4.0
	DAY	**RAʰ**	**DEC°**	**MAG**
Mars	5	14.7	S17	+ 0.1
	15	15.0	S19	+ 0.2
	25	15.4	S21	+ 0.3
Jupiter	15	2.2	N12	− 2.5
Saturn	15	3.0	N15	+ 0.3

September

Moon	Last 2	New 9		
	First 17	Full 25		
Venus	AM			− 4.5
	DAY	**RAʰ**	**DEC°**	**MAG**
Mars	5	15.9	S22	+ 0.3
	15	16.4	S23	+ 0.4
	25	16.9	S24	+ 0.5
Jupiter	15	2.2	N12	− 2.8
Saturn	15	3.0	N14	+ 0.1

October

Moon	Last 2	New 9		
	First 17	Full 24		
	Last 31			
Venus	AM			− 4.5
	DAY	**RAʰ**	**DEC°**	**MAG**
Mars	5	17.4	S25	+ 0.5
	15	17.9	S25	+ 0.6
	25	18.4	S25	+ 0.7
Jupiter	15	2.0	N10	− 2.9
Saturn	15	2.9	N14	− 0.1

November

Moon	New 8	First 16		
	Full 23	Last 29		
Venus	AM			− 4.3
	DAY	**RAʰ**	**DEC°**	**MAG**
Mars	5	19.0	S24	+ 0.8
	15	19.6	S23	+ 0.8
	25	20.1	S20	+ 0.9
Jupiter	15	1.7	N9	− 2.9
Saturn	15	2.8	N13	+ 0.2

December

Moon	New 7	First 16		
	Full 22	Last 29		
Venus	AM			− 4.1
	DAY	**RAʰ**	**DEC°**	**MAG**
Mars	5	20.6	S20	+ 0.9
	15	21.1	S18	+ 1.0
	25	21.7	S15	+ 1.0
Jupiter	15	1.6	N8	− 2.7
Saturn	15	2.6	N13	0

Pollux 7.8ʰ N28°, Regulus 10.1ʰ N12°, Spica 13.4ʰ S11°, Antares 16.5ʰ S26°

Appendix B: Calendar for the Moon and Four Planets (continued)

2000

January

Moon	New 6	First 14	
	Full 21	Last 28	
Venus	AM		−4.0

	DAY	RAʰ	DEC°	MAG
Mars	5	22.2	S12	+1.1
	15	22.7	S9	+1.1
	25	23.2	S6	+1.2
Jupiter	15	1.6	N9	−2.4
Saturn	15	2.6	N13	+0.2

February

Moon	New 5	First 12	
	Full 19	Last 27	
Venus	AM		−4.0

	DAY	RAʰ	DEC°	MAG
Mars	5	23.7	S3	+1.2
	15	0.1	0	+1.2
	25	0.6	S3	+1.3
Jupiter	15	1.9	N10	−2.2
Saturn	15	2.6	N13	+0.3

March

Moon	New 6	First 13	
	Full 20	Last 28	
Venus	AM		−3.9

	DAY	RAʰ	DEC°	MAG
Mars	5	1.0	N6	+1.3
	15	1.5	N9	+1.4
	25	1.9	N12	+1.4
Jupiter	15	2.2	N12	−2.1
Saturn	15	2.8	N14	+0.3

April

Moon	New 4	First 11	
	Full 18	Last 26	
Venus	AM		−3.9

	DAY	RAʰ	DEC°	MAG
Mars	5	2.5	N15	+1.4
	15	2.9	N17	+1.5
	25	3.4	N19	+1.5
Jupiter	15	2.7	N15	−2.0
		-o- after 24		
Saturn	15	3.0	N15	+0.3

May

Moon	New 4	First 10	
	Full 18	Last 26	
Venus	AM to 6		−3.9

	DAY	RAʰ	DEC°	MAG
Mars	5	3.9	N21	+1.5
	15	4.4	N22	+1.5
		-o- after 24		
Jupiter		-o- to 22		
Saturn		-o- to 22		

June

Moon	New 2	First 9	
	Full 16	Last 25	
Venus		-o-	

	DAY	RAʰ	DEC°	MAG
Mars		-o- all month		
Jupiter	15	3.6	N19	−2.1
Saturn	15	3.5	N17	+0.3

Nearby reference stars—all magnitude about 1.2: Aldebaran 4.6ʰ N17°,

Appendix B: Calendar for the Moon and Four Planets (continued)

2000

July

Moon	New 1	First 8		
	Full 16	Last 24		
	New 31			

Venus	PM after 20			− 3.9
	DAY	**RA**[h]	**DEC°**	**MAG**
Mars		-o all month		
Jupiter	15	4.1	N20	− 2.1
Saturn	15	3.7	N18	+ 0.3

August

Moon	First 7	Full 15		
	Last 22	New 29		

Venus	PM			− 3.9
	DAY	**RA**[h]	**DEC°**	**MAG**
Mars		-o to 6		
	15	8.8	N19	+ 1.8
	25	9.2	N17	+ 1.8
Jupiter	15	4.4	N21	− 2.3
Saturn	15	3.9	N18	+ 0.2

September

Moon	First 5	Full 13		
	Last 21	New 27		

Venus	PM			− 3.9
	DAY	**RA**[h]	**DEC°**	**MAG**
Mars	5	9.7	N15	+ 1.8
	15	10.1	N13	+ 1.8
	25	10.5	N11	+ 1.8
Jupiter	15	4.6	N21	− 2.5
Saturn	15	4.0	N18	+ 0.1

October

Moon	First 5	Full 13		
	Last 20	New 27		

Venus	PM			− 4.0
	DAY	**RA**[h]	**DEC°**	**MAG**
Mars	5	10.9	N8	+ 1.8
	15	11.3	N6	+ 1.8
	25	11.6	N4	+ 1.8
Jupiter	15	4.6	N21	− 2.7
Saturn	15	3.9	N18	− 0.1

November

Moon	First 4	Full 11		
	Last 18	New 25		

Venus	PM			− 3.5
	DAY	**RA**[h]	**DEC°**	**MAG**
Mars	5	12.1	N1	+ 1.7
	15	12.4	S3	+ 1.7
	25	12.8	S4	+ 1.7
Jupiter	15	4.4	N21	− 2.8
Saturn	15	3.7	N17	− 0.3

December

Moon	First 4	Full 11		
	Full 18	New 25		

Venus	PM			− 4.2
	DAY	**RA**[h]	**DEC°**	**MAG**
Mars	5	13.2	S6	+ 1.6
	15	13.6	S8	+ 1.5
	25	14.0	S11	+ 1.4
Jupiter	15	4.1	N20	− 2.8
Saturn	15	3.6	N17	− 0.2

Pollux 7.8[h] *N28°, Regulus 10.1*[h] *N12°, Spica 13.4*[h] *S11°, Antares 16.5*[h] *S26°*

Appendix C: Periods of Visibility for the Planet Mercury with Magnitude

	Before Sunrise	After Sunset
1990	Jan 21 (+0.5) to Mar 7 (−0.8) May 24 (+1.2) to Jun 25 (−1.5) Sep 18 (+0.5) to Oct 8 (−1.1)	Mar 29 (−1.5) to Apr 15 (+0.5) Jul 11 (−1.1) to Aug 14 (+0.5) Nov 8 (−0.6) to Dec 15 (+0.4)
1991	Jan 4 (+0.4) to Feb 18 (−0.8) May 14 (+0.5) to Jun 8 (−1.2) Sep 5 (0.4) to Sep 22 (−1.2) Dec 18 (+0.4) to Dec 31 (−0.3)	Mar 13 (−1.4) to Mar 30 (+0.4) Jun 26 (−1.2) to Jul 25 (+0.5) Oct 19 (−0.6) to Nov 28 (+0.5)
1992	Jan 1 (−0.3) to Jan 18 (−0.4) Apr 23 (+0.5) to May 23 (−1.2) Aug 19 (+0.5) to Sep 5 (−1.4) Dec 2 (+0.2) to Dec 31 (−0.5)	Feb 24 (−1.3) to Mar 14 (+0.5) Jun 9 (−1.3) to Jul 4 (+0.5) Sep 28 (−0.7) to Nov 10 (+0.4)
1993	Jan 1 (−0.5) to Jan 5 (−0.6) Mar 19 (+1.9) to May 5 (−1.1) Jul 27 (+1.7) to Aug 18 (−1.3) Nov 12 (+1.5) to Dec 14 (−0.6)	Feb 7 (−1.2) to Feb 27 (+0.8) May 25 (−1.3) to Jun 30 (+1.7) Sep 11 (−0.7) to Oct 31 (+1.9)
1994	Feb 28 (+2.0) to Apr 20 (−0.7) Jul 7 (+1.8) to Aug 3 (−1.3) Oct 27 (+1.5) to Nov 24 (−0.7)	Jan 21 (−1.0) to Feb 14 (+1.0) May 11 (−1.2) to Jun 12 (+1.8) Aug 26 (−0.6) to Oct 15 (+1.9)
1995	Feb 11 (+1.6) to Apr 3 (−0.8) Jun 21 (+1.6) to Jul 19 (−1.4) Oct 12 (+1.6) to Nov 7 (−0.9)	Jan 2 (−0.8) to Jan 28 (+1.3) Apr 24 (−1.4) to May 22 (+1.8) Aug 7 (−1.0) to Sep 27 (+1.9) Dec 12 (−0.7) to Dec 31 (−0.6)
1996	Jan 26 (+1.6) to Mar 16 (−0.8) May 31 (+1.8) to Jul 1 (−1.1) Sep 26 (+1.5) to Oct 17 (−1.0)	Jan 1 (−0.6) to Jan 13 (+1.6) Apr 8 (−1.3) to May 2 (+1.8) Jul 21 (−1.0) to Sep 8 (+1.8) Nov 22 (−0.6) to Dec 28 (+1.9)

Appendix C: Periods of Visibility for the Planet Mercury (continued)

	Before Sunrise	After Sunset
1997	Jan 8 (+ 1.8) to Feb 26 (− 0.7) May 9 (+ 1.9) to Jun 17 (− 1.3) Sep 8 (+ 1.8) to Sep 30 (− 1.2) Dec 24 (+ 1.4) to Dec 31 (0.0)	Mar 23 (− 1.3) to Apr 14 (+ 1.9) Jul 7 (− 0.9) to Aug 22 (+ 1.9) Oct 31 (− 0.6) to Dec 12 (+ 1.8)
1998	Jan 1 (0.0) to Feb 7 (− 0.7) Apr 19 (+ 1.9) to Jun 1 (− 1.2) Aug 23 (+ 1.9) to Sep 14 (− 1.3) Dec 7 (+ 1.7) to Dec 31 (− 0.4)	Mar 7 (− 1.4) to Mar 28 (+ 1.7) Jun 19 (− 1.3) to Aug 2 (+ 1.9) Oct 10 (− 0.6) to Nov 26 (+ 1.8)
1999	Jan 1 (− 0.4) to Jan 17 (− 0.6) Mar 31 (+ 1.9) to May 16 (− 1.2) Aug 6 (+ 2.0) to Aug 29 (− 1.3) Nov 21 (+ 2.0) to Dec 28 (− 0.6)	Feb 19 (− 1.2) to Mar 10 (+ 1.2) Jun 4 (− 1.4) to Jul 14 (+ 2.0) Sep 22 (− 0.7) to Nov 9 (+ 1.7)
2000	Mar 11 (+ 1.8) to Apr 28 (− 1.0) Jul 18 (+ 2.0) to Aug 12 (− 1.3) Nov 5 (+ 1.8) to Dec 4 (− 0.7)	Feb 1 (− 1.1) to Feb 24 (+ 2.0) May 18 (− 1.3) to Jun 20 (+ 1.7) Sep 3 (− 0.8) to Oct 24 (+ 1.9)
2001	Feb 18 (+ 1.7) to Apr 12 (− 0.9) Jun 23 (+ 2.0) to Jul 26 (− 1.3) Oct 15 (+ 1.1) to Nov 13 (− 0.8)	Jan 1 (− 0.8) to Feb 8 (+ 1.7) May 1 (− 1.4) to Jun 6 (+ 2.0) Aug 15 (− 0.8) to Oct 8 (+ 1.8) Dec 22 (− 0.6) to Dec 31 (− 0.6)
2002	Feb 3 (+ 1.5) to Mar 24 (− 0.6) Jun 5 (+ 1.5) to Jul 11 (− 1.3) Oct 3 (+ 1.5) to Oct 26 (− 0.9)	Jan 1 (− 0.6) to Jan 21 (+ 1.2) Apr 17 (− 1.1) to May 16 (+ 2.0) Jul 30 (− 0.9) to Sep 20 (+ 2.0) Dec 1 (− 0.5) to Dec 31 (+ 0.3)
2003	Jan 23 (+ 0.5) to Mar 9 (− 0.8) May 26 (+ 1.2) to Jun 25 (− 1.5) Sep 20 (+ 0.5) to Oct 10 (− 1.1)	Mar 31 (− 1.5) to Apr 17 (+ 0.5) Jul 13 (− 1.1) to Aug 16 (+ 0.5) Nov 10 (− 0.6) to Dec 17 (+ 0.4)

Appendix D: Modern Climate Zones

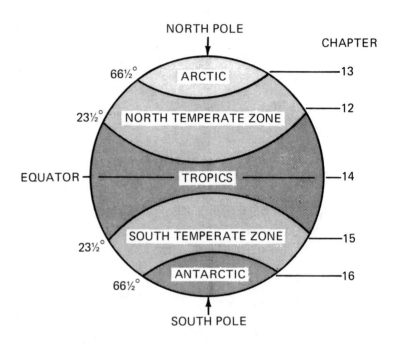

Appendix E: Brightest Stars and Their Constellations

#	Star Name	MAG	Constellation	RA^h	DEC°
5	Achernar	0.6	Eridanus	1.6	S57 s
30	Acrux	1.1	Crux	12.4	S63 s
10	Aldebaran	1.1	Taurus	4.6	N17
51	Altair	0.9	Aquila	19.8	N9
42	Antares	1.2	Scorpius	16.5	S26
37	Arcturus	0.2	Bootes	14.3	N19
—	Becrux	1.4	Crux	12.8	S60 s
16	Betelgeuse	0.5[1]	Orion	5.9	N7
17	Canopus	– 0.9	Carina	6.4	S53 s
12	Capella	0.2	Auriga	5.3	N46
53	Deneb	1.3	Cygnus	20.7	N45
56	Fomalhaut	1.3	Piscis Austr.	23.0	S30
31	Gacrux	1.6[2]	Crux	12.5	S57 s
35	Hadar	0.9	Centaurus	14.0	S60 s
21	Pollux	1.2	Gemini	7.7	N28
20	Procyon	0.5	Canis Minor	7.6	N5
26	Regulus	1.3	Leo	10.1	N12
11	Rigel	0.3	Orion	5.2	S8
38	Rigil Kent	0.1	Centaurus	14.7	S61 s
18	Sirius	– 1.6	Canis Major	6.8	S17
33	Spica	1.2	Virgo	13.4	S11
49	Vega	0.1	Lyra	18.6	N39

#—*the navigational star number (see Chapter 9).*

s—*southern star, not visible north of about latitude 35°N.*

[1]—*mean magnitude; Betelgeuse's magnitude varies from 0.1 to 1.2.*

[2]—*Gacrux just misses being a first-magnitude star; somewhat brighter, nearby Becrux is not a navigational star.*

Appendix F: Key to Pronunciation

The stressed syllable is printed in SMALL capital letters.
Here is the key for the vowel sounds:

a	as in fat	ay	as in fate
e	as in wet	ee	as in feet
i	as in it	eye	as in ice
o	as in odd	oh	as in go
uh	as in up	you	as in use
		oo	as in food

Notice that all the sounds in the left column are short. In the right column they're long and like the names of the letters a, e, i, o, and u.

The "uh" symbol also stands for the neutral sound in unstressed syllables in words like sofa, kitten, pencil, lemon, and circus. Followed by the letter r the "uh" will also do for the sounds in earn, irk, and urn. You don't need a special symbol for "ar"; sound it as in arm.

Latin	Pronunciation	English
Aries	AR-eez or AR-ee-ez	Ram
Taurus	TO-ruhs	Bull
Gemini	JEM-i-neye or -nee	Twins
Cancer	KAN-suhr	Crab
Leo	LEE-oh	Lion
Virgo	VUHR-goh	Virgin
Libra	LEE-bruh	Scales
Scorpius	SKOR-pi-uhs	Scorpion
Sagittarius	saj-i-TAY-ree-uhs	Archer
Capricornus	kap-ree-KOR-nuhs	Sea Goat
Aquarius	uh-KWAYR-ee-uhs	Water Carrier
Pisces	PEYE-seez	Fishes

Index

Page numbers for figures are in boldface type.